Something of the Marvelous

Lessons Learned
From Nature and
My Sixty Years
as an Environmentalist

Huey D. Johnson

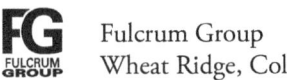

Fulcrum Group
Wheat Ridge, Colorado

Copyright © 2020 Resource Renewal Institute

All rights reserved. No part of this book may be reproduced or transmitted in any form or by any means electronic or mechanical, including photocopying, recording, or by any information storage and retrieval system except by a reviewer who may quote brief passages in a review without permission in writing from the publisher.

All photos are from the collection of the Resource Renewal Institute unless otherwise noted.

Cover woodcut illustrations by Fiona G. King.
Cover design by Chance Cutrano.

ISBN: 978-1-56373-200-3
Library of Congress Control Number: 2020947689

0 9 8 7 6 5 4 3 2 1

Fulcrum Group
3970 Youngfield Street
Wheat Ridge, CO 80033
303-277-1623
www.FulcrumGroupBooks.com

To my wonderful family,
who have been kind enough to join me on the journey of my life:
Sue, whose support and adventurous spirit
has made everything possible;
Megan, Tyler, and Jill, who are my closest friends and
outstanding environmentalists in their own right;
and Miles and Bay, already accomplished young men
and the first audience for my book.

Top row: my son Tyler, and grandsons Miles and Bay.
Bottom row: my daughter Megan, wife Sue, daughter-in-law
Jill, and Bochy and Buffy.

In all things of nature
there is something of the marvelous.
— Aristotle

What else can we do in life that would possibly
be here a thousand years from now?
— Huey Johnson

Contents

- ix   Foreword
  by Barbara Deutsch
- xiii Acknowledgments
  by Deborah Moskowitz
- xvii Introduction

1	Chapter One	Lesson Learned:	How to Tie a Knot
11	Chapter Two	Lesson Learned:	Know When to Walk Away
20	Chapter Three	Lesson Learned:	Travel Solo and You'll Never Be Alone
37	Chapter Four	Lesson Learned:	Find Your Drum
55	Chapter Five	Lesson Learned:	A Little Craziness Is Good
93	Chapter Six	Lesson Learned:	Never Give Up. Never Give In.
120	Chapter Seven	Lesson Learned:	Drink Forty Cups of Tea
147	Chapter Eight	Lesson Learned:	Be a General and a Generalist
194	Chapter Nine	Lesson Learned:	Think Big. Act Bigger.
215	Chapter Ten	Lesson Learned:	Living the Lessons
266	Conclusion	Lesson Learned:	Follow Your Nature

- 271 Bibliography
- 273 Index
- 283 About the Author

Foreword

He led the charge to save Hawaii's Pools of 'Ohe'o (Seven Sacred Pools) for posterity. He wrested mile after mile of Bay Area coastline from imminent development. He spearheaded the protection of twelve thousand miles of wild rivers and more than six million acres of wilderness and open land. He received the prestigious United Nations environmental prize. Along the way, he crossed paths with legendary figures like David Brower and Wangari Maathai. Ansel Adams and Wallace Stegner. Jerry Brown and Phillip Burton. William O. Douglas and Margaret Mead.

Sadly, environmentalist Huey Johnson passed away in July 2020 at age eighty-seven just a few weeks after completing his memoir. Happily, he led a life full of adventure and accomplishment. For Huey, a life experience was only as valuable as the lesson he learned from it, a lesson that made the next experience more valuable, and the next one after that. Even more important to

Huey was sharing those lessons with others, especially younger people like the dozens of environmentalists he took under his wing throughout his career, many of whom have gone on to lead their own environmental organizations and projects. As one of his former mentees explains, "Anyone who works for Huey should actually pay tuition."

Huey learned his life lessons from writers, thinkers, and colleagues on the one hand and from his own bold and independent choices on the other. As a young man, he left a promising corporate career to travel the world, educating himself in history, art, culture – and human nature. He abandoned a PhD program at the University of Michigan to head west and find his true calling, having little idea what that might be. Chancing upon a book by Aldo Leopold on a Lake Tahoe bookshelf, he found "his drum" and set out to become an environmentalist.

With a belief in the power of generalism and the wisdom of hiring experts to fill in the rest, he became the first Nature Conservancy director in the West, then created The Trust for Public Land, an organization that innovated ingenious methods for acquiring and preserving open space in urban settings. As California secretary of resources in the first Brown administration, Huey engaged with allies and rivals alike to protect California's unique environment from exploitation and degradation, finding win-win solutions when possible and adamantly standing his ground when necessary.

In 1985, Huey founded the Resource Renewal Institute as his sandbox, a grassroots nonprofit where he could put into action the lessons learned throughout his career. At RRI, he and a passionate band of colleagues, volunteers, and advisors put together a series of inventive, effective programs whose far-ranging influence belies RRI's tiny size.

The lessons in this book are both thought-provoking and practical. When to compromise and when to hang tough. How a little madness can lead to great things. The difference between love and respect. The lifelong value of acquiring basic skills.

Not surprisingly, Huey found many of his most meaningful lessons in nature itself, using his own inner resources to survive life-threatening encounters with bears, glacial ice, and roaring rivers. He continued to hike, fish, and hunt into his eighties, revitalizing his commitment to the natural world with every outing.

People who have worked with Huey say that anyone who met him immediately became an environmentalist. Here is your chance to get to know him in his own words, and if you're not an environmentalist already, to become one yourself.

Barbara Deutsch
Environmental writer
and creative consultant

Acknowledgments

In 2012 I started working with Huey Johnson at Resource Renewal Institute, the environmental nonprofit he founded in 1985. As it happened, it was exactly fifty years after a twenty-seven-year-old Huey first read *A Sand County Almanac* and was so inspired by Aldo Leopold's land ethic that it set the course for his life's work. Huey's successful land saving and resource policy accomplishments are so vast, they are difficult to enumerate. It's harder still to comprehend how he got it all done.

In early 2018 Huey and I began exploring the idea of a book that would frame his collection of stories and case studies in the context of "lessons learned," an approach he felt was more relevant and useful to others than a memoir. A daunting task loomed: to organize a three-inch pile of essays and remembrances that he'd been writing and rewriting for at least six years, and weave them into book form. Huey recognized that this job required an entirely different skill set from his own.

Fortunately, the ideal person to tackle this task was already well known to us – writer Barbara Deutsch, who had completed several key projects for Resource Renewal Institute. Barbara is an empathetic listener and expert wordsmith with an extraordinary ability to synthesize complex information and present it in a creative and compelling fashion. When Huey asked her to help, she enthusiastically stepped up, embraced

the "lessons learned" vision for the book, and came up with a plan (and a process) to get it done.

A generous grant provided by Marion French Rockefeller Weber, Huey's precious longtime friend, provided the initial support for Resource Renewal Institute to hire Barbara and take on Huey's book as a project. As the work progressed, something wonderful happened to us at Resource Renewal. The chance to hear Huey's life stories again, to relive his remarkable optimism in the face of adversity, provided us with an uplifting respite from the environmental challenges and troubling political news of the day.

With Huey and Barbara working in close collaboration, *Something of the Marvelous* took shape and came to life. Huey was so pleased with a first draft of the manuscript that he surprised us by sending it off to his good friend Bob Baron at Fulcrum Publishing. Bob offered to take the Fourth of July weekend to read it and let Huey know what he thought.

A few days later, Huey fell and suffered a serious, and ultimately fatal injury. He spent a week in the hospital and then returned home to his family. As I drove over to visit Huey on Sunday, July 5, Bob called me to say that Fulcrum would like to publish his book, adding, "A book is special. It is personal and it can last. An author can write words that transcend time and place. Aldo Leopold and *A Sand County Almanac* tremendously influenced Huey as a young man. Huey's words and ideas will influence others in the coming years. It is part of his legacy." I was able to deliver this wonderful message to Huey, who squeezed my hand and said, "Huzzah."

With guidance, encouragement, and an exceptional commitment from Bob Baron and Fulcrum's talented staff, creative director Patty Maher and senior editor Alison Auch, we have been able to carry Huey's book across the finish line in far less time than we ever imagined. Thanks also to Chance Cutrano, director of special programs at Resource Renewal Institute, for his striking cover design featuring Huey's beloved sandhill crane, inspired by an earlier collaboration between graphic designer Wayne Kosaka and woodblock illustrator Fiona G. King.

And, especially to Sue Johnson, Huey's wife and partner of nearly sixty years, and to his family, our heartfelt appreciation for their help in completing his book during a very difficult time.

Huey demonstrated the ability to learn from life's experiences and to use those lessons learned for self-improvement – continually raising the bar for what he could achieve. Even in his eighties, he never let the challenges of age stop him from pursuing his work and outdoor adventures. Instead, he calmly assessed his limitations and worked out practical solutions that allowed him to keep doing what he loved, albeit in somewhat different ways. As he often said, "Don't let the perfect be the enemy of the good." Huey passed along an extraordinary example of a life lived to its fullest – tirelessly, ethically, and joyfully.

We miss Huey terribly but are grateful for all he has left for us. His commitment to making the world a better place is an inspiration to all of us. His lesson of perseverance is a call to each of us to continue his work.

Huey often quoted to me from his favorite poem, *Ulysses*, by Alfred Lord Tennyson. Its last words capture him well.

Made weak by time and fate, but strong in will
To strive, to seek, to find, and not to yield.

Deborah Moskowitz
President, Resource Renewal Institute
August 2020

Introduction

There was no way down.

I was alone on a boulder, out of rifle shells, and surrounded by ice. Paralyzed with fear, soaked in sweat, I thought this must be what it feels like to face a firing squad.

It wasn't the smartest thing I'd ever done, heading off on my own that morning. But I was twenty-four years old, staying at a youth hostel in New Zealand, and I felt like hunting for thar, a wild mountain goat with curved horns and delicious meat. It never occurred to me to ask someone to join me or to tell the hostel staff where I was going.

I had climbed above the snow line and begun crossing a glacier. After a while, I saw a thar in the distance. I stalked and shot him, but he slid down the glacier and vaulted off a cliff into space. A moment later, the thar's fate was nearly my own as I slipped and slid off a glacier myself. I dropped about twenty feet, landing on a large boulder that was lodged precariously in the ice. I was uninjured, but I was stranded thousands of vertical feet from the ground below. I had only about two feet of boulder surface to move around on and was far enough from where I had fallen that there was no way to get back up.

I sat on the boulder for a long time. Every escape plan I could think of was too dangerous to try. But eventually, I concluded that if I stayed where I was, I was going to die. I had to do something. I had to act.

Putting the rifle sling around my neck, I looked for a spot in the boulder to grab with my hands so I could descend to the slick glacial ice below. Not a pleasant prospect, but it had to be done. As I dangled over the edge, the boulder broke free from its resting point in the ice. I started to fall, but my hands caught the lip of a depression that the boulder had formed in the ice. I hung there, getting my wits about me. Then I inched my way around and got back to the point where the boulder had sat a few minutes before. I studied the rocks on either side of my resting place.

I saw cracks in the rock wall where I could hinge my fingers, and I launched forth. I had probably gone several hundred yards when I came to a rockslide path polished slick by past slides. Below was a sheer ice cliff, thousands of feet high. It was impossible to go any farther, but somehow I did. I slid into a chute – which should have been the end of me, but my foot happened to catch on a crack, and I was able to edge my way across the chute.

Looking back at the places I had just traversed, I was overcome with a sense of relief and a belief that this was not my day to die. A calm settled over me, a mystical clearing of fear. I felt cool and confident, now quite comfortable taking extreme risks. I jumped easily from point to point, like a monkey in the jungle. As I made my way down, I gleefully teetered on pointed rocks without using my hands at all.

There were still challenges ahead, but I met them with new resolve. Facing another vertical ice wall, I used my rifle barrel as an ax and carved steps I could walk on for several hundred yards. That gave me the idea to use the rifle barrel on rock face too, creating divots to ease my descent. The rifle – now with a bent barrel – would never again do what it was intended to, but it helped me survive when I needed it.

As the sun lowered on the horizon, it lit up the mountain slopes. At last, I could see the bottom and the safety it represented. I noticed a golden glint on the rocks. As I got closer, I realized it was the thar I had killed earlier that day. And although not many hours had passed since I shot the animal, it had been an eternity in life experience and what I had learned about myself.

I begin this book about my life with the story of the thar because it represents my belief that a well-lived life is one filled with lessons learned. Sixty years ago on that icy mountain, I had learned a lot. Choose life over fear. Use whatever tools you have to survive, especially your own mind. Remember that you are more resourceful than you think. And never again forget to prepare ahead for the awesome power of nature.

In my opinion, there are three different kinds of life lessons: the ones you learn yourself – like my mountain adventure; the ones you learn from others; and the ones you learn from great books. My life has been rich in all three. And I have found that as these lessons enhance and build on one another, life grows continually more interesting and satisfying.

Sharing my love of nature with Megan and Tyler.

My hope is that this book offers readers some lessons of the second kind – ones learned from others. In this case, from me. Lessons like how to travel the world on your own and on a budget. When to stand your ground and when to compromise. How to turn your back on a life that's wrong for you and discover the life you should be living. How to fight hard but learn from your adversaries. How to find renewal and sustenance in nature. And how to tie a knot.

I often think about the leather-bound books that line the shelves of law firms. They contain centuries of legal knowledge, landmark lawsuits, and historic precedents, providing attorneys with the tools they need to make widespread social change through the courts. I think environmentalists need something similar, an archive of the methods and strategies that successful activists have used to make groundbreaking environmental change – so the next generation won't have to start at square one every time they take on a new challenge.

I want to start that ball rolling. That's why my book includes so many stories of how I and the people I worked with over the years have found ways to make sweeping environmental change. Like establishing the United Nations Environment Programme. Sharing the concept of Greenplanning with governments around the world. And preserving millions of acres of wilderness as public lands, including Hawaii's Seven Sacred Pools and twelve hundred miles of wild California rivers and streams. In every case, there was a method to our madness, a method that may help others in their work to reverse climate change, protect our oceans, or preserve wildlife habitat.

CHAPTER ONE

Lesson Learned | How to Tie a Knot

I think learning is life's most precious asset, and growing up in rural Michigan, I was blessed with a rich and varied education. Very little of what I learned, however, took place in the schools I attended.

Instead, I learned the things that aren't taught in school. How to fix motors and broken fences. How to hunt and fish for my protein. How to feel connected to nature. How to survive in the wilderness when I have to. How to feel confident and self-reliant in the world. And how to tie a knot that will hold.

If I felt short-changed by my inadequate formal education, it turned out to have a positive effect on me. I discovered a lifelong

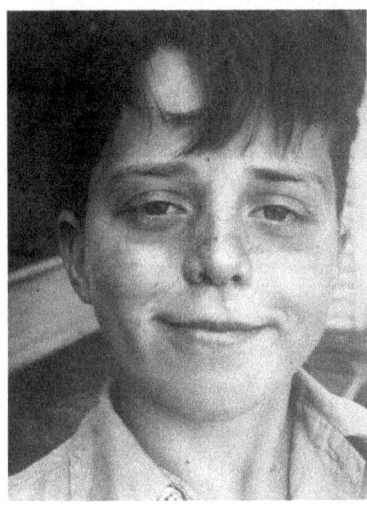

Growing up in rural Michigan.

passion for reading and books, and for teaching myself whatever I needed or wanted to learn. From an early age, I haunted the small Carnegie Library in our town, often waiting patiently for their mail to arrive so I could read the latest news magazines. It's no wonder I have always felt a special appreciation for Andrew Carnegie, the industrial tycoon and philanthropist, who established 1,946 libraries in America's small rural communities between 1883 and 1929. If it weren't for his ideals and inspiration, who knows what I would have become.

At our Carnegie Library, I also developed a habit I have continued throughout my life: I read something useful and informative for at least an hour every day. It's a practice that has served me remarkably well in my life and career, allowing me to grasp the major issues and salient ideas I needed to act on as an environmentalist, political activist, and manager of thousands of government employees.

When I was growing up, the Depression was coming to a close, and although my parents had limited finances, we were well fed, healthy, and able to live in our own home. Their families had lived in Michigan for many generations. My great-grandfather on my father's side was an immigrant from Sweden. The clerks at Ellis Island changed his last name from Swenson to the more American-sounding Johnson. My mother's ancestors came to the States from France. They were descended from the Huguenots, the French protestants who were persecuted for centuries. I don't know when her family immigrated to America, but it was early enough that some of them fought in the Civil War. I have a French middle name, Dernier, which means "the last." I was to be their last child – until my younger sister was born.

Tragically, one of the family members who died before I was born was not from the distant past. My four-year-old brother, Duane, was killed one year before I was born when a car went out of control, jumped the curb, and hit him while he was riding his tricycle. My parents

I was raised to be an independent child.

were shattered and remained so for years. Among my earliest memories is their weeping in each other's arms.

In addition to the brother I never knew, I had two delightful sisters. Rebecca was four years older than I, and Marilyn four years younger. As adults, they both chose careers that helped others, which always made me proud. After Rebecca raised her family, she became a teacher in an underserved community and helped develop a free tuition program that gave hundreds of students a chance for a college education. Marilyn was a gifted and compassionate caregiver in a retirement home.

Duane's death and my parents' grief could have cast a long shadow over my childhood, but for some reason, it only helped me thrive. My parents made it clear that I was expected to live for both myself and my brother, and to always try harder in my endeavors. They often said as much when issuing some work directive or commending or criticizing me. It was a tough way to motivate a child, but I knew my parents loved me, and the responsibility they put on my shoulders benefited me in my growth. I always wanted to fulfill their dream, and I think I did. They lived to see me become a member of Governor Jerry Brown's cabinet as secretary of the California Department of Resources. The newspaper clippings and scrapbooks they collected were evidence of how proud they were of who and what I had become.

Throughout their lives, my parents valued honor, obligation, and, above all, hard work. As far back as I can remember, I always had some kind of job, and without my even thinking about it, I developed a strong work ethic. As a young boy, I gardened, landscaped, stocked shelves, and delivered newspapers – which led to my short-lived career as an unwitting juvenile delinquent. One day when I was six or seven, I was doing my usual job of delivering papers on my bicycle to the local stores that sold them to the public. I noticed that people were giving money to a man in military uniform who gave them paper flowers in return, red poppies to be exact. Of course, the man was an American Legion volunteer raising funds for World War I veterans. I, however, knew nothing of that. Enterprising youngster that I was, I came up with a plan. I would take the money I earned from my paper route, buy the man's paper flowers and resell them on the street at a profit. Which is just what I did. I "bought" all the man's poppies for a dime each and then sold them on the street for 50 cents apiece. That evening, I proudly showed my proceeds for the day to my father, who promptly marched me back downtown to return the money to the American Legion volunteer.

At age fifteen, I graduated to painting bridges and then to a relatively high-paying assembly line job at an auto factory. Sometimes I worked the night shift, and when my friends and I got off work, we would drive out to the lake and swim. I worked through grade school and high school and college, and with little time to study, my grades were only average. It wasn't until I went to grad school and my wife was generous enough to support us that I finally had my days free to study, and I excelled academically.

I didn't just acquire new skills and proficiencies from my jobs as a child. I also learned about life and adults and relating to people from different backgrounds, which set me apart from my friends with more sheltered upbringings. One of my most memorable role models was Fred Murray, the owner of a shoe store where I stocked shelves when I was eight years old. Mr. Murray was one of the most respected members of our small community, and his integrity and humanity left a deep impression on me. Though his store was a humble affair, he dressed like a New York banker in a three-piece suit complete with a flower in the lapel. While he chatted with the customer seated before him, I would run and get the requested footwear from the stockroom. His customers were principally poor farmers who were as likely as not to have manure on their boots, yet he treated them like kings. He did the same with everyone who entered his store – always kind, respectful, and smiling broadly. Most of all, he was honest and honorable, qualities in short supply both then and now.

Even with my responsibilities, there was always time for nature. I have no doubt that I became an environmentalist because of my parents' love and reverence for the natural world. I remember going on wildflower walks with my mother when I was a toddler. And she often took my sister and me walking in a park with a vigorous pine forest that she had helped plant when she was a young child herself. My father hunted and fished, and as I grew old enough, I joined him on his adventures. As early as the first grade, my friends and I were independent enough to hike out to a camp our fathers helped us build on a stream a few miles from town. As little fishermen, we learned precious life lessons: how to be quiet and patient, and how to think like a trout so we could anticipate its behavior and location. At dusk, one of our fathers would show up to make sure we were all right for the night. Otherwise we were on our own. By the time we were nine and ten, our little group had become ice fishers and rabbit hunters and had taught our dogs how to hunt. Never supervised, we managed just fine.

It's only in retrospect that I find myself surprised at the level of responsibility, self-sufficiency, and independence that defined my childhood. In addition to our jobs, we planted, picked, and canned our vegetables, and raised and killed chickens and rabbits. The adults suggested we not give names to the animals so we wouldn't feel too close to them. That was the extent of our sentimental attachment to the creatures we lived with.

Guns were an integral and normal part of my life from an early age. I was taught that my father's short-range shotguns were deadly, and was trusted to act accordingly. He did remove the shells from his guns when they weren't in use, but beyond that, it was my responsibility to know when, where, how, and why to shoot.

To be honest, my little band of friends and I weren't always so responsible. Our long hours of independence sometimes got us into some mischief. For some reason, we liked to hang around a nearby sauerkraut factory. We would run around the machinery, making lots of noise and driving the workers crazy. One of them threatened us again and again that if we didn't stop our shenanigans, he would dunk us in a vat of sauerkraut. We didn't believe him, of course, until he did just that.

When I became a teenager, I encountered a challenge that all my hunting, reading, and factory work hadn't prepared me for – the opposite sex.

Although we had moved to Lansing when I was in the ninth grade, my friends and I still enjoyed going to Saturday night country dances at some remote rural four corners dance hall. I would call my wonderful uncle Charley, a well-known fiddle player in those parts, and he would clue me in on where there was going to be a dance.

At one dance, I met a lovely young lady and we proceeded to enjoy the evening together. When the dance was over, the party moved to a nearby farm where we continued dancing and partying into the early morning hours. It was about four a.m. when I finally drove her home to the farm where she lived with her parents. It was a hot and muggy Michigan August night, and the farmhouse's screened windows and doors were all open to keep out insects and let in air. Trying not to wake anyone, we tiptoed to the front door where I nervously tested my resolve to kiss her goodnight.

Just as I got to the crucial moment, there was a deafening roar from inside the house. The door shot open and a huge man leaped out, shouting and slamming the door – and me – against the wall of the porch. He had me pinned. With about six inches and the screen door between us, he laid into me: "Nobody brings my daughter home at four in the morning…" Then suddenly his voice became gentle and he smiled widely, "…unless he stays for breakfast and helps me milk the cows." So I milked cows for the first time, had a delicious breakfast, and rolled out of there about nine a.m. with a sweet memory and another lesson learned: How to take a joke.

After high school, I moved on to Michigan State University in East Lansing. Like Utah State University where I got my master's degree several years later, MSU is a public land grant school, which meant that, to some extent, I have Abraham Lincoln to thank for my education. In a farsighted and underappreciated action, Lincoln signed the Morrill Act of 1862. The law gave federally controlled land to each state free of charge, with the mandate that it either establish a college on the property or sell the land and use the proceeds to build a college elsewhere in the state. Responding to the social changes brought on by the Industrial Revolution, the law required these institutions to focus on practical subjects like science, engineering, and agriculture. The goal was to make college affordable and available to a far wider population than the children of privilege who attended the nation's largely private liberal arts schools. In other words, people like me. That's why I have always had a special place in my heart for President Lincoln. Without him, I might never have had the excellent college and graduate education that opened up so much of the world to me.

Of course, I still had to work my way through school, both at MSU, and later, at Western Michigan University in Kalamazoo, where I transferred to and received my bachelor's degree in biology. For a while, I was the evening desk clerk at the YWCA. It was an ideal job for a student because the front door was locked at eight o'clock, and I could work and study at the same time. As a young college boy employed by a women's hotel, I knew some of the people I encountered were suspicious of my motives. I made a point of being especially polite and responsible.

My major challenge was in the dining room. As it got close to closing time, I was charged with peeling the women guests away from their male dinner dates and sending the gentlemen on their way. It only became a problem when couples had had too much to drink and were determined to go upstairs together. Separating the women from their muscle-bound companions could be daunting, and whenever I managed to get a guest into the elevator alone, I felt the satisfaction of a job well done.

In the summers, I was lucky enough to be a member of a garbage collecting crew. The sanitation department hired extra men in the summer because, in those days, homemakers often preserved food during the harvest season, filling up enormous hundred-pound containers of wet, heavy peelings from tomatoes and peaches. Our job was to lift or drag the containers onto the garbage trucks.

Why did I consider myself lucky to be a garbage man, a job I imagine most people would reject out of hand? Because our trucks rolled out of a large collection center at five a.m., and, regardless of how long it took to finish our daily route, we were paid for eight hours – and paid very well. So we worked like mad dogs, running with the large empty cans and picking up the full ones as fast as we could. Usually, we were finished in four hours. I was in the best physical shape of my life.

Most of us took the opportunity to get a second eight-hour job. Between the two, we saved enough money to pay for our next college year. That's why the garbage collection job was so sought after. Those of us who did it earned an extra dose of pride.

Another summer, I was a Fuller Brush man with a rural territory. I went from farm to farm, knocking on doors. Back before Amazon and shopping malls, it was a great convenience for rural families to order brushes and other cleaning supplies out of their homes. I would take their orders and deliver the products a few weeks later. I was paid on a commission basis and did well at it.

I acquired such invaluable skills as fending off any large, vicious guard dog who rushed at my car. By deftly inserting my sample case between the dog and my leg, I could get out of the car and maneuver myself to the farmhouse. I learned that I should always go to the back door because everyone on a farm was working outdoors or in the kitchen.

I could write an entire book about my adventures as a Fuller Brush man. One evening it was nearly dark, but as a good salesman, I was determined to make one more call before ending my day. I saw a light coming from a forest beside the road. I followed it to a large house trailer and knocked on the door. The door was opened by an attractive young woman wearing only her underwear. She invited me in. Although it was too dark for her to see my wares, she already knew what she wanted, so we sat on the couch while I took her large order for cosmetics.

Boom! I heard a crash. A huge man jumped into the trailer, screaming, "Aha!" and, "You SOB, I'm getting my shotgun." He rifled through a nearby closet and came out with an armload of stuff, including his gun. As I hurriedly finished packing up my samples, I saw that the hulking husband was planted directly between me and the trailer door.

Recalling my football training, I ran at him and piled into him as hard as I could. He went down in a heap. I stepped on him on my way out. I never came back with the order.

There was one time in college when the skills I was acquiring as a salesman did more than just provide me with a livelihood. They actually saved my life. Two buddies and I were driving back from an outing of some kind when we stopped to give a hitchhiker a ride. He got in the car, said hello, then pulled out a handgun and stuck it in my face. It was the most terrifying thing that's ever happened to me, before or since. I realized at that moment that if I wanted to come out of this alive, I would have to *sell* this guy on not killing us. I talked and talked, setting out my reasons for why he shouldn't pull the trigger. I told him he could take our possessions, so what good would it do to kill us and wind up in prison, we wouldn't tell anyone what happened, we were just some young kids with our lives ahead of us, and on and on and on. I was convinced, rightly so as it turned out, that if I just kept talking, he wouldn't shoot. I still can't quite believe it, but my sales pitch for my own life was successful. I closed the deal.

In my twenties and thirties, I went through periods of feeling insecure about my education, especially as I crossed paths with graduates of the Ivy League schools I had idealized since I was a child. But the more I worked with these people, the more I realized that I was their equal in every way, that I had had a great education too. From the hundreds of books and newspapers I pored over at the Carnegie Library to the wild assortment of jobs I took to pay for my schooling to the years I spent traveling and absorbing world art, history, and culture, I had become a truly educated man. Even more important, I became someone who had the habit of learning, a habit that pleases and enriches me every day of my life.

CHAPTER TWO

Lesson Learned | Know When to Walk Away

It was Friday evening. I was sitting at a Manhattan bar, waiting for my roommates to join me so we could launch the weekend with an evening of fun. Reaching for my cocktail, I noticed my hand was shaking. That was new. I thought about it. I'd been working hard for several weeks with no days off. There had been no time to get myself out into nature, which was my way to relax and unwind. I was still in my mid-twenties, but I was a nervous wreck. Would I get ulcers next?

I had chosen the corporate life with enthusiasm. I studied biology and marketing at Western Michigan University with the intention of moving up in the world. I wanted to be rich and successful, and as luck would have it, I was offered a well-paying job that made use of both my college majors. My employer was Visking, now known as Viskase and owned by Union Carbide. The founder had invented a cellulose material that food companies used to encase hot dogs and other processed foods. Before his invention, frankfurters had always been encased in animal intestines, a tedious procedure that didn't work nearly as well as cellulose. It might not seem like a major invention, but when you consider the number of hot dogs consumed every day, and the fact that each casing could be used only once, it's not surprising that Visking was a major, highly profitable corporation. In fact, the demand for our casings was so great that materials were delivered by the boxcar – literally.

| *A Visking sales meeting.*

I was part of a team of about twenty technical salespeople known as "service sellers" who were tasked with keeping the business rolling and our food-packing customers well-serviced and happy. I traveled regularly from my office in New York to branch offices and customer sites in Chicago, Minneapolis, Denver, Boise, and other locations. The job was considered technical because we had specialized knowledge that allowed us to correct problems and make adjustments to our customers' sensitive automated, high-speed machinery. Most of the time, I showed up at a customer's factory every few weeks to check out their operation and prevent emergencies. I could also be called on at a moment's notice to go anywhere in the States when an assembly line stopped working. That was serious business because every minute an operation was off-line, big money was going down the drain.

The founder of Visking may have invented cellulose casing, but there were soon many competitors in the field. It was up to us service sellers to maintain our company's industry leadership by visiting and entertaining our customers regularly, making sure they continued to use hundreds of tons of *our* cellulose each week. I usually took factory owners or buyers and their wives to dinner. Occasionally we went waterskiing, took fishing trips, and spent evenings at the theater. My customers in the West were often in remote locations hundreds of miles apart. Because they didn't see new faces too often, I made it a point to be well-read and provide interesting dinner conversation. And although it was part of my job to cultivate these relationships, many of these clients became real friends.

There was much to love about my job. I worked hard, enjoyed my independence, made my sales quotas, and learned how to manage my time and responsibilities. All in all, I seemed to have a knack for the salesman's life. Unlike my colleagues with families, I was single and fancy-free so I didn't experience the pain they felt being away from home for days and weeks at a time. Happily, my efforts were recognized and rewarded with raises and promotions. I won a prestigious national sales contest no one of my age and relative inexperience had ever won before. I had been advised not to even bother entering the competition because there was no way I could win. When I did win, it was a pivotal moment in my life, an official confirmation that I had the ability to succeed, that I could make my way in this world. I was infused with confidence. After all, a good salesman could always get a job. I wasn't going to starve.

The best development of all was being assigned to the Rocky Mountain territory headquartered in Denver. I flew into Denver the next day, rented a car, and drove up into the mountains. I immediately fell in love with the western way of life. My first assignment was a two-week swing through west Texas where I spent the weekend horseback riding in Bandera, known as the dude ranch capital of the world. Saturday night, we feasted on cabrito, young goat roasted over an open fire. Another assignment took me to San Francisco for training. After seeing the city for the first time, I wondered why anyone would ever live anywhere else. Back in Denver, I arranged my car so it held my skis, guns, fishing tackle, and outdoor gear. I made friends wherever I went and spent every weekend enjoying myself in a different beautiful place – skiing in Sun Valley or Jackson, fishing for steelhead in Oregon streams, reliving Hemingway's bird-hunting experiences in Idaho.

Being a Visking cellulose service seller had become a dream job. Then I was transferred back to New York.

I had been living in Colorado for about six months when I received a call on a Friday night. I was told I would be returning to New York permanently, starting the following Monday. The person I was replacing had died in a horrible accident. He was running to catch an elevated train, missed the last step onto the train car, and fell onto the tracks. I had seen him just a few weeks earlier at a staff meeting where he hugged me and enthused about how lucky we were to work for such a "nice, caring company that was like having another mother."

Arriving on Monday I emptied the dead man's desk and took the box with his things to his home in New Jersey. I knocked on the door of his small house, and his weeping widow invited me in. The company her husband had described as "another mother" hadn't bothered to send flowers or condolences. I stammered my sorrow and left. From that moment on, I took corporate loyalty with a grain of salt.

I did love the vitality and diversity of New York. I especially enjoyed servicing our four hundred kosher accounts, meat processors who sold to New York's large Jewish community. Many of these clients made me feel part of the family, inviting me to weddings, bar mitzvahs, and other celebrations.

Still, the intensity of living in New York and the pressures of my job began to mount, and it wasn't long before I found myself sitting in that Manhattan bar with shaking hands and almost no life to call my own. I began to think seriously of leaving Visking.

The truth is, I had begun to experience doubts about Visking even before I was transferred to New York. Some of it had to do with my own innate sense of independence – or streak of rebellion, depending on your point of view. Some of it, paradoxically, had to do with the fact that the company had opened my eyes to a wider world, which made me want to do more with my life. And some of it had to do with the company itself.

For example, IBM was the paragon of corporate discipline at that time, and Visking followed the IBM dress code of gray suit, white shirt, and red tie. Newly installed at the Denver headquarters, I admired how the local ranchers dressed – in a light-colored suit topped off by a big-brimmed Stetson hat. I decided to wear just that to our next national sales meeting; I was the only one there who broke the corporate dress code. I took some serious ribbing about my clothes, but I never got into trouble. I produced for the company so they left me alone.

Another thorn in my side was the officious head of the accounting department. He required the sales force to feed him endless sheets of statistical project information. He saw himself as our superior. My view was just the opposite: his role was to support our sales effort. He complained to the big boss, which led to a meeting where I made it clear that I could either spend my time filling out forms or I could do what I was paid to do – sell our company wares. I was never asked to fill out another form.

I was also infuriated by our lavish expense accounts. When I was new to the company, my spending on the road reflected the way I was raised and had always lived. Frugally. Assuming the company wanted me to watch their pennies, I usually ate my meals at diners or low-cost cafés and stayed at budget hotels or motels. After a few months on the job, I was called in to the big boss's office. He closed his door and sat me down.

"Why are you eating so little for dinner?"

I explained that I was eating just fine at reasonably priced restaurants. He expressed the same criticism about where I stayed. I said I was trying to hold down my costs, assuming it was company policy.

"No," he said, "that's not what I mean. Our clients judge us by where we stay and where we have dinner. Plus, the sales force gets some of their income from their expense accounts. Don't turn in any more forms like these." He returned my recent expense forms to me to be redone.

"Your starving yourself risks the expense accounts for the rest of the sales force who turn in high ones. I want you to eat steak dinner in nice places. And if you aren't hungry, then eat two steaks the next night. And you should always try to entertain a client and his wife for dinner."

It was a lesson in padding an expense account, and from that day on, I dined exclusively in the finest establishments of Chicago, Denver, and New York City.

Perhaps my moment of greatest doubt came when I was standing outside of a huge factory that used our products. At about the third-story level, I saw a chute spewing the tattered remains of the stuff I sold to that firm to aid their manufacturing. I watched it become a waste pile as large as a house. Something didn't feel right.

And so, after nearly three years at Visking, I decided to resign. I would forgo my goal of becoming wealthy. I would no longer impress dates with dinners of steak and champagne. I would give up the ego gratification that went along with the praise of my superiors, prestigious speaking engagements, and especially, my high salary. In my opinion, I was lucky to have achieved material success at such a young age because I learned it had no meaning for me. I discovered that I was actually an idealist, that I wanted a life that had something to do with nature. What that "something" was I still didn't know.

Once my mind was made up, it still took courage to resign. I was in Colorado at the time and sent a one-sentence telegram to company management tendering my resignation. There was an uproar from the home office. I was instructed to return at once to corporate headquarters in Chicago to talk things over. I thanked them but declined, whereupon my immediate superior flew out to Denver and pleaded with me to stay. He tried to make me feel guilty by telling me that upper management held him responsible for my leaving, and he feared for his job. He succeeded. I did feel guilty and went to Chicago to talk to the big boss.

Facing the boss in person was a fearsome thing. It almost never happened unless you were going to be reprimanded, fired, or promoted. I entered a huge room where at least a hundred employees were seated at their desks. All eyes were on me as I took the long walk to the other end and reached the imposing oak door that led into the boss's terrifying office. Every element of the experience was designed to intimidate me. It did.

But I had a few psychological tricks up my sleeve too. My plan was to head off on a round-the-world adventure. To keep myself from being seduced into staying with the company, I had purchased my airline ticket in advance.

I knocked on the door.

"Come in," came the lion's roar from behind the massive door. As I approached his desk, there was silence. I noticed how big his fists were as they clenched on the desktop. He motioned for me to sit down. He rose, walked over to the door, and closed it.

"Johnson, what is this crap that you're going to quit?"

I managed to squeak out, "I have quit, sir."

His response: "Don't you know we've poured a lot of money in your training, and we have very high expectations for you?"

"Thank you for that," I said. After glaring at me for a while, he turned away from the desk and stood, silently looking out the window at the belching smokestacks of his industrial complex. An eternity passed.

Then he turned, walked over, and loomed over me.

"You're serious about this, aren't you?"

Afraid of what I would sound like if I spoke, I nodded my head yes and pulled the round-the-world plane ticket from my pocket. To my astonishment, he stuck out his hand, grasped mine, and shook it.

"You lucky bastard," he said. And with one movement of his hand, he gestured that I should go and waved good-bye.

He had given me a wonderful tonic – freedom.

I walked past his oak door into the large room of employees. All eyes were on me again. This time, I was smiling with joy and relief and immediately went home to pack. At that point in my life, it was the most important decision I had ever made. When I think of everything it led to afterward, it may have been the most important decision I made in my entire life.

Before I got on the plane, however, there were still more temptations not to go, and they weren't easy to resist. The first was from the owner of a very successful company that supplied machinery to Visking. He took me to lunch and said, "Huey, I want to retire. I don't have children or an heir. I want to leave the company to you to run as your own. I've watched you and I know you would do it well."

The second offer came from a meatpacking plant in Idaho, a place where I would have liked to live. I'll always remember the sight of the owner, standing out by his stock pens in the ice and snow.

"Huey, I have a son I hoped would take over from me. He told me he doesn't want to do it. He wants to be a doctor. I'd like to offer you the business. I've worked with you for the last few years and know you would do it well."

These were extraordinary opportunities for a young man – not one but two chances to own an established business with no financial investment on my part. I thought hard, but I had already decided that becoming wealthy wasn't important to me anymore, that the business world wasn't my destiny. And I was determined to use my trip around the world to try and figure out what that destiny was.

I never regretted working at Visking. How could I? I became an efficient manager of my own time. I saw enough of the world to know I wanted to see more. I gained knowledge and acquired skills that have served me throughout my environmental career, especially the value of salesmanship – being a good listener, building rapport, presenting my case strategically and persuasively, anticipating a client's concerns. Best of all, Visking was where I found out that I could compete, succeed, and often excel in the adult world. It was there that I developed confidence in myself.

Perhaps it was that growing confidence that made it possible for me to learn one of life's most valuable lessons: Know when to walk away. I've put that lesson to use many other times in my life. But that was the first and the hardest. And I never looked back.

My beloved San Francisco Bay.

CHAPTER THREE

Lesson Learned | Travel Solo and You'll Never Be Alone

What set me off on the idea of traveling around the world with no plans, no itinerary, no companions, and no idea how long I would be away?

Not many Americans did such a thing in 1960, and travel wasn't as easy as it is today. Airlines were just introducing jet planes. There were no ATMs or email. One overseas phone call could cost hundreds of dollars. Money was sent by wire, and letters were delivered to a local American Express office after weeks in transit. That worked for big cities only. If you were staying in a small town, you had to rely on the local postal service, which could take a month or two to deliver a letter or package. On the other hand, it was the era of *Europe on Five Dollars a Day*, and the American dollar was strong as an ox. I started out with $3,500, knowing it would support me for quite a while.

I went overseas to discover myself as much as the world. I hoped to find out what interested and excited me so I could pursue a career that had meaning and contributed something to others. Two and a half years at Visking had convinced me I was not cut out for big business. I had an abiding love of nature and was familiar with great visionaries like John Muir and David Brower, but the idea of turning my passion for the wilderness into a career was unimaginable to me.

I had always had an intense curiosity about the world, whetted by the books I read at the Carnegie Library growing up and the Visking sales trips that introduced me to the grandeur of the western United States. I wanted to see the earth's natural wonders and understand its history, art, and myriad cultures. I was especially interested in exploring two subjects. One was religion. I'd fallen away from the intense religious customs of my small-town upbringing, but I was still curious about the different ways people around the globe expressed their faith. My other area of interest was how natural disasters and environmental mismanagement influenced history and transformed world civilizations.

The people I knew and respected were baffled by my choice to leave my promising career at Visking. My parents were worried and disappointed. But my mind was made up, and the moment the plane took off, I experienced a joy I'd never known before. It came from the realization that I had literally years ahead of me with no schedule, no expectations, and no responsibility to anyone but myself.

The most important lesson I learned in over two years on the road was to travel alone, the very thing that most of us are loathe to do. When you're on your own, you have no choice but to engage with others, to meet people who are also by themselves. All at once, you get to know people from other countries and cultures, something that just doesn't happen when you're with friends or family. Not only do you make new friends, you gain real-time tips from travelers who've just visited a place where you intend to go. "You're off to Singapore? I stayed at a great hostel for 50 cents a night. And be sure you go to the end of the street for dumplings. The owner's name is James. Say hi for me." Best of all, you'll always have friends to visit in countries you've never set foot in before.

But "travel alone" was by no means the only lesson I learned in my two and a half years away. A few more:

Make no plans.
Taking the time to stay somewhere that fascinates you, or conversely, leaving immediately when you've had enough, is a delicious luxury. So is the pleasure of simply sitting on a park bench observing people or reading a book for as long as you please. So when you can, avoid taking tours or buying tickets and hotel rooms in advance. See what happens and follow your instincts.

Don't fly from one place to the next.
Hitchhike or take a ship or train, and you'll create more opportunities to engage with new people and places.

Always go cheap.
Stay at hostels, college dorms, monasteries, and YMCAs or rent a room in someone's home, Eat at cafeterias and steam tables, or get takeout from

grocery stores and cafés. Again, it's the best way to meet the locals, make new friends, and have fun.

Pack a suit and tie (or the equivalent if you're a woman).
It was astonishing how an ordinary suit and tie opened doors for me when I was traveling. Whenever I showed up at a US embassy or a national holiday celebration wearing that traditional uniform, I was immediately welcomed in for gourmet dinners, potent cocktails, enchanting musical performances, and other delights that rarely came my way when I was wearing jeans and a T-shirt. When going by ship, I was invariably ushered directly to the coveted captain's dining table where we were served specially prepared dishes and the finest wines. With attendance by invitation only, the captain's guests included sophisticated, often noteworthy people from around the world who had mastered the art of sparkling conversation. I ate up every bite – and every word.

Celebrating South Africa's independence with the crew at Tongariro National Park in New Zealand.

It never mattered that I hadn't been invited beforehand or that no one knew who the hell I was. My suit seemed to communicate that I would behave myself reasonably well and would be at least mildly interesting company. Back then, Americans working in foreign capitals rarely came in contact with American travelers. It made them eager to meet someone new, someone from back home. It was particularly true of younger embassy employees who were desperate to meet young people they could have fun with in the evenings. They had access to liquor at a greatly discounted price so after the older people went to bed, we headed for their private social rooms and had a blast. The icing on the cake: I could usually stay overnight for free in the quarters of the Marines who guarded our embassies.

And it all came from wearing a suit and tie.

I have no idea if our embassies are as welcoming today as they were then, but I'm convinced my suit-and-tie strategy will still give you an entrée to many wonderful places where a T-shirt and jeans just won't do.

Another travel secret I found invaluable was my membership in the Freemasons. I can honestly say that I would never have had the trip I did were it not for the Masons who befriended me in almost every country I visited. A remarkable fraternal organization with millions of members worldwide, Freemasonry has a seven-hundred-year history steeped in tradition, symbolism, and ritual. Fourteen US presidents, many Supreme Court justices, Winston Churchill, Simon de Bolivar, and leaders of the Russian Revolution were all members. The Masons remain a secret society based on a foundation of an unassailable trust that its members have for one another.

Every Masonic lodge I visited embraced me as a fellow member, often taking me home and including me in memorable local experiences. I happened to be in New Delhi, for example, where the Masons were holding their annual international convention. I showed up, validated my membership with a few words and gestures known only to Masons, and joined the group. Addressing the crowd was a tall, imposing man whose pink head wrap and rich blue jacket identified him as a follower of the Sikh religion. He stood next to a table where holy books from every major religion were spread open for viewing. They illustrated his point that the Masons' valued the wisdom of all religions and welcomed members from all faiths. Before long, someone I met invited me to his country club for some fishing. It was like no fishing trip I'd ever experienced. We floated on a raft fitted with plush upholstered chairs. When I idly dipped my fingers in the lake, I was warned that I might become the next meal for the crocodiles swimming below.

I found equally warm welcomes and unique adventures whenever and wherever I stopped into a Masonic lodge. It was like a golden key to the world.

After deciding to start out in Asia, I wandered across the Pacific and soon found myself in New Zealand. Even after a lifetime of world travel, I still think New Zealand is the most beautiful place I've ever been. It was there that I began to finally slow down from the go-go salesman's life. Which led me to cash in my plane ticket and stay put for a while. That's when I learned yet another travel lesson. I could stretch my money as well as my trip by working along the way.

I got a job at a ski resort. It happened by accident when I was hitchhiking on the North Island and noted a snow-capped mountain on the horizon. My ride described it as Tongariro National Park, a majestic place with three active volcanoes, alpine trails, emerald lakes, boiling mud pools, and the Tongariro River with its plentiful trout and rapids for rafting. He added that there was also a famous ski resort in the park.

I immediately asked him to drop me off, then I hitched another ride to the top of the mountain. I wandered around a bit, walked into the lodge, and applied for a job. To my astonishment. I was hired as a member of the mountain crew, the team that kept the ski runs and lifts functioning. I didn't have any cold weather gear with me, however. The problem was easily solved by hitchhiking to the next town where I found a used-clothing shop and quickly outfitted myself with the appropriate gear. Then I hitched back up to the ski resort and joined my new coworkers, friendly young people who represented fourteen different nationalities.

When spring ended the ski season, I became a professional hunter for about a month. I was hitchhiking on the South Island, and some men in an old truck picked me up. They were freelance hunters paid by the New Zealand government to kill the red tail deer that populated the surrounding mountains – at $3.50 "a tail." They told me that New Zealand was overrun with large ruminants because early settlers had imported them from Scotland (red deer), the European Alps (chamois), and the Himalayas (thar). With no native predators, their numbers had exploded, causing harmful erosion from overgrazing. The hunters had an extra rifle and asked me to join them in the precipitous glacial mountain terrain. Before long, I had shot a chamois and a thar. It was later that same day that I recklessly headed out on my own to do some more hunting, a life-changing experience I describe in the Introduction to this book.

I guess I proved my marksmanship because the hunters invited me to share their cabin on beautiful Lake Wakatipu and take part in their daily excursions.

Thinking back, I realize that neither of these unforgettable job opportunities would have come my way if I hadn't been a hitchhiker. Thumbing rides connected me with real New Zealanders who often invited me for dinner and welcomed me to stay overnight in their homes. None of it would have happened if I had stayed on the typical tourist trail.

In spite of all the hunting I had done in New Zealand, I still had one unfulfilled dream: to go on a guided hunting expedition in this exotic land. I could never afford to pay for it, of course, so I stopped by a guide service located in a sporting goods store and offered to do anything to earn my keep on one of their luxury expeditions.

Out of the blue, the wealthy store owner invited me to his elegant mansion for the evening. (Once again, I was glad I had packed my suit and tie.) Over dinner, he made me an astonishing offer. He would take me on the hunting trip of a lifetime – no cost, no work, no strings attached.

It was such a generous gesture, I just couldn't get it through my head that it was genuine, that he meant what he said. I felt embarrassed, which led me to hem and haw: "I sure don't want to be any trouble to you in any way, sir." Suddenly he interrupted me. "Forget it. If you don't trust me, I don't trust you." And that was it. The hunting trip I dreamed of was canceled just seconds after it was offered.

Once I left his home and got over my shock, I thought a lot about what he said. I began to understand something I've remembered the rest of my life. *Trust generosity.* Let someone be nice to you. Appreciate your good fortune. Realize that you have something to offer other than money. You can have wonderful experiences and become friends with many delightful people that way. Even more important, experiencing other people's generosity teaches you to be generous in turn.

Trust generosity. It was one the most valuable lessons I learned on my trip. And in my life.

Most of my New Zealand adventures occurred out-of-doors, but I'll never forgot one that happened indoors – at the Museum of New Zealand in Wellington. Strolling into one of the museum's exhibition rooms, I was amazed by the imposing sight of a chair-sized block of pure green jade.

It was truly a mystical experience. I later found out that the jade piece was a shrine to a god worshipped by the Maori, the indigenous people of New Zealand who arrived by canoe from eastern Polynesia around the fourteenth century. Later, at the Wellington Zoo, I was equally moved to hold a live kiwi in my hand. Kiwis are the endangered flightless birds whose numbers were decimated by the cats that arrived aboard the ships that brought European settlers.

After about four months in New Zealand, I boarded a ship for Australia. As I tried to imagine what Australia would be like, I remembered something I hadn't thought about since I was a small child. In an elementary school class, the teacher asked us all to stand up and tell everyone what we wanted to do when we grew up. While other children said they wanted to be firemen or teachers or president of the United States, I answered: "I want to go to Australia with a quarter in my pocket." Recalling that moment, I thought I must have already had wanderlust as a child, even if I didn't understand where or what Australia was. And now I was about to make that childhood dream come true.

When I arrived in Australia, I started looking for a job. I got hired as a temporary sales manager for a chain of advertising agencies with offices across the country. Why did I take a job so similar to the one I had gone halfway around the world to escape? Because it gave me the chance to explore most of Australia – a country nearly as large as the United States – with all expenses paid.

I knew that Visking had an operation in Australia, so I stopped by a supermarket one day to see which local products were encased with Visking materials. Having spent a good chunk of my life ensuring that those casings were used properly, I was shocked to see how shoddy and unappetizing the Visking-packaged products looked on those Australian supermarket shelves. I knew immediately that Visking had to be losing a pile of money. I found out they had hired a British company to run their

Australian operation. They were running it all right – into the ground. I wrote a long letter to Visking management in America to let them know what was happening. Then and there, they offered me ownership of the Australian business. It was an opportunity that represented certain wealth for the rest of my life in a country I had come to love. Thinking on the offer, I realized it was a test of how far I had come, both literally and figuratively, from the corporate life I abandoned months before. The answer was I had come a long, long way. I turned the business down without a thought, continuing on with my life of adventure.

The exotic natural environment of Australia fascinated me, and I took every opportunity to explore it. I spent hour upon hour at the Royal Botanic Garden in Sydney. I went hiking with a researcher who was studying the duck-billed platypus, a protected endangered species nearly hunted to extinction for its fur. We watched them frolicking in a natural pool and saw several deadly snakes along the trail.

By this point, I was an established member of the international travelers' subculture with connections everywhere I went. I soon moved on to Singapore where I contacted Chinese friends I had met in universities in Australia. I was treated generously. One of my friends got me a temporary membership in his Chinese country club. After a while, a Caucasian friend asked if I would come to his wedding in Kuala Lumpur in Malaysia. His wife was Chinese, and he didn't have any relatives or other friends who would be attending. I had a wonderful time. He and I were the only Caucasians in attendance, and I felt so fortunate to be part of a large traditional Chinese wedding.

Traveling alone continued to pay dividends. So did being single. At the wedding, I met a young woman who was a student nurse on Penang Island in the South China Sea, a few hundred miles to the north. She invited me to visit, and I went. The island had a Shangri La–like quality, with a high plateau covered in jungle vegetation and a tram you could take to the mountaintop. The student nurses treated me well. We often had picnics on a beautiful sand beach or went to the top of the plateau for a hike. We ate at wonderful Chinese restaurants where the food was so cheap, I could afford to be generous. After several weeks, I said good-bye to my new friends in Penang and caught a train to Bangkok. There was a guerilla uprising in the region, and in every train car, armed soldiers stared out of the windows, looking for insurgents.

In Singapore, Kuala Lumpur, Penang, Thailand, and Burma, the religious architecture was magnificent. I knew little about Buddhism when I got off the boat in Singapore, but I read a great deal about it while I toured around Southeast Asia. In Thailand and Burma, I saw giant statues of Buddha and other wondrous carvings and explored mesmerizing temples where snakes crawled on every surface and hung from the ceilings.

Every rule has its exception, including my rule about traveling without plans. For me, the exception was India. How I wish I had prepared myself for what was to come. When I arrived in Calcutta, the shock was instant. I saw beauty and sorrow in every direction, and felt elation and sadness at every turn. I wandered in India for a few months, winding up in Bombay (now Mumbai). I believed I would never be the same after my experience in this complex, contradictory country, and I never was.

My intention was to get a low-cost ticket on a ship from Bombay to Europe. I checked in to the local YMCA and encountered a number of people with the same goal. Each day we would check on ship arrivals to see if there were any low-cost fares back to Europe. Finally, Mario and I, a Spaniard I had met at the Y, scored tickets for the same ship to Marseilles. I paid for his ticket based on his assurance that money would be waiting for him when we landed. We arrived in Marseille, but the money didn't. Mario called his family who suggested we somehow get ourselves to their home in Barcelona, where I would be reimbursed – not a bad prospect at all. I bought an old Vespa motor scooter, and we were off to Spain.

We had barely crossed the French-Spanish border when someone hailed us from a parked car. It was the mayor of a village a few miles down the road who proposed a banquet in our honor. Why in the world would he do such a thing? Because he was a close friend of Mario's father, a high-ranking officer in the Spanish Republican Army during the Spanish Civil War who was still considered an enemy by the current Fascist regime. Low on money and living on a loaf of bread a day, we couldn't believe our good fortune. I still remember the wonderful evening. The meal was superb, and the wine a deep blood-like red that was the best I ever tasted.

As we continued our ride to Barcelona, I saw people carrying large bags of what I soon found out were pesetas, the currency of Spain before the euro. The irresponsible economic practices of the Franco dictatorship had created runaway inflation, and the population needed more and more money each day to pay for what they bought – unless, of course, they made their purchases on the black market, which is exactly what almost everyone did. Along the way, we were often stopped at roadblocks by Franco's soldiers who pointed guns at us and searched our bodies and belongings. Obviously, the reputation of Mario's father preceded us wherever we went.

In Barcelona, Mario introduced me to his friends, many of whom were intensely opposed to Franco. We frequented a Republican club below street level that we entered through the back room of a bar. I assume Franco's forces knew about the club, but as long as we stayed underground, they didn't interfere.

Barcelona also marked a milestone for me, the most important of my life. I met an American girl, Sue Madden, who was working as a guide at the World's Fair in Brussels. In time, we were married, and now, more than fifty years later, are still happily hitched with a fine family.

I left Mario, Barcelona, and the girl who would become my wife to explore Europe by Vespa, eventually covering eight thousand miles. I became obsessed with visiting the great museums where Western history, culture, and civilization converged. I was a sponge, enthralled and excited by everything I saw. I honestly believe I learned more at these magnificent treasuries of knowledge than I did in all the years I spent in high school, college, and graduate school.

At that time, I had a lot to learn. I remember visiting Saint Peter's Basilica in Rome, standing among all the tourists and observing the works assembled in that handsome building. I came to a place where a small marble statue was drawing a lot of interest from those around me. I was moved by its perfection, frozen in place and time like a bird dog on point.

"What is that?" I whispered to the person next to me. "It's the *Pieta*. Michelangelo's *Pieta!*" I had never heard of it or him.

I decided to stay in Rome for a while to immerse myself in Italian Renaissance art and history – just as I had done in Southeast Asia when I became fascinated with Buddhism. With no plans or schedule – and the luxury of total freedom – I dived into the churches, museums, palazzos, and piazzas of Rome, watching and reading as much as I could absorb.

In Paris, I soaked up the Louvre, Notre Dame, and the other famous museums and sites. My favorite masterpiece was, and still is, the *Winged Victory*, sculpted in the second century BC and an inspiration to everyone who lays eyes on its soaring beauty. One day, while standing in line to pick up my mail at the Paris American Express office, I was lucky enough to meet two young American women who were art students. They were spending the summer studying Impressionism in Paris, and we soon became fast friends. They guided me through the many museums in Paris that featured Impressionist work. I absorbed a lot on the subject in the week I spent with them, and I still love and study Impressionist painting.

On a side note: more than fifty years later, I visited the Hermitage Museum in St. Petersburg Russia, whose six buildings make up the largest art museum in the world. There are many galleries of Impressionist paintings on exhibit – perhaps even more than in Paris. It was a glorious display, and I felt a moment of gratitude for those young graduate students who began my lifelong love affair with Impressionism half a century before.

My favorite place in Europe was Westminster Abbey in London. I moved slowly through that immense structure covered with statuary reflecting England's heritage. I wandered off into a side area and ended up in what looked like a small church apart from the main structure. The space was silent. A few beams of sunlight shone through the stained glass windows and landed on the pews below.

I sat for a time, then stood and turned to go. Just then, I noticed a man in black robes standing in a shadow nearby. I said hello. He smiled and asked if I knew where I was (he could spot a small-town boy). I said no. He pointed to the floor where I saw the word "Tennyson" etched in the stone. Without knowing it, I was standing on the grave of the great poet, Alfred Lord Tennyson. The man in black waved his hand slowly at the floor and walls. It finally dawned on me that I was at Poet's Corner – the resting place for England's great literary figures. He pointed to a headstone in the wall engraved with "Geoffrey Chaucer – 1343–1400." I sat down again, only this time I studied the floor and walls. Over there were the Browning sisters, next to them the composer Handel. I have been back to Poet's Corner a number of times since, but the first time – when I didn't know what to expect – was the most thrilling of all.

My experiences as an innocent abroad planted seeds that have borne fruit throughout my life and career. Learning about Buddhism in Southeast Asia led me to donate a West Marin County ranch I acquired at The Nature Conservancy to the San Francisco Zen Center. It became Green Gulch, a renowned mecca for Zen practice, education, and organic farming. My commitment to saving the Seven Sacred Pools in Maui was informed by my exposure to the beautiful indigenous cultures of New Zealand and Australia. The peace and joy I found in the city parks throughout Asia and Europe were instrumental in my founding The Trust for Public Land, a nonprofit that acquires and protects open spaces in urban settings. Even my visit to a New Zealand used-clothing store changed me forever. Since that day, I have never bought a new piece of clothing, and I imagine I've saved many thousands of dollars over the years.

In the bigger picture, even as I was nourished by the cultural riches of Europe, Africa, and Asia, I came to appreciate that the United States has a magnificent heritage no Old World country can match – our majestic wilderness. Protected by forebears of great vision, these millions of acres of national parks, public lands, and wildlife refuges are our "old masters" – the wondrous sites that attract visitors from around the world to our country.

Lesson learned: We Americans must guard our precious heritage of wilderness as zealously as the museums of older civilizations guard their masterpieces.

I have one last story from my time in Europe, and it's a beaut. I was heading into Hamburg, Germany, on my Vespa. Going down a steep hill at full speed, I came around a bend and saw a traffic jam directly in front of me. In my lane was a lineup of enormous military trucks. Unable to stop in time, I had no choice but to overturn the scooter on its side, then slide under the truck rather than hit it head-on. There was an awful crunching sound. My helmet bounced on the cement like a ball, leaving me all but wrapped around the back of the truck ahead of me.

I remember hearing someone say, "Is he dead?" which I would have been if not for the cocoon-like structure of my Vespa's front panel. Crouching behind the curved piece of metal, I was protected from the worst of the impact. People pulled the scooter free and worked me out from under the truck. I had a broken bone but was able to get the scooter running well enough to ride into Hamburg for medical treatment.

Outfitted with a cast, I spent my days in an American library doing my favorite thing – reading. I got to know an attractive young woman who worked at the checkout desk. Though German, she was a graduate of an American University.

We were soon dating, and she took me home to meet her parents. They appeared to be prosperous. My relationship with their daughter progressed, and it wasn't long before they invited me to use their guest room when I came to visit. One night it became clear that my girlfriend wanted to talk about our relationship.

"Since we're getting serious, I have something to tell you about my future. My father was a high-ranking officer in the army. We still believe in Hitler's leadership. I want to have children to serve the next Third Reich."

I said goodnight, collected my pack, walked away, and never talked to her again.

The timing of our breakup was perfect. After two and a half years on the road, I had finally had my fill of eating out, getting lost, staying in hostels, missing my family and friends, and having no place to call my own. I was ready to come home.

Across Asia, the Pacific, and Europe, I had been dazzled and enriched by the treasures I encountered. I left home with the narrowest grasp of the natural and cultural heritage of our planet. With every visit to a museum, historic place, archaeological site, or natural wonder, my understanding widened and deepened. The trip also stirred thoughts about the arc of history and the effect of human behavior on our environment. But looking back, it is the unforgettable people I met that I treasure most – the exciting duality of how different the world's cultures can be but how much we still have in common as human beings.

My worldwide adventure had been glorious and made me the person I am today. As I had hoped, it pointed me in the direction of how I wanted to spend the rest of my life. It was the best thing I've ever done.

CHAPTER FOUR

Lesson Learned | Find Your Drum

My trip around the world may have been over, but my journey to find my life's work was just beginning. It took another couple of years – in Colorado, California, Alaska, Idaho, Utah, and Michigan – to make that happen.

While I was overseas, I spent a lot of time visiting battlegrounds and reading about war and social collapse throughout history. After a while, I realized that a single culprit was responsible for many troubles – the human mismanagement of natural resources. I walked the arid land of the Sahara Desert. Once a rich agricultural area, it had been desiccated thousands of years ago by overgrazing. I visited the sites where ancient Romans clear-cut their forests and overirrigated their crops, practices that caused salting of the soil. The excess of salt prevented wheat from growing, resulting in widespread hunger and food riots that contributed to the empire's decline. These experiences led me to read authors like one-time Michigan governor Chase Osborne and the visionary ecologist Paul Sears, whose 1935 book *Deserts on the March* describes how the mismanagement of topsoil has had disastrous results throughout history – including the Dust Bowl of the Great Depression.

I thought about getting into a career that focused on these issues, though I wasn't sure what that would be. Science? Politics? The Park Service? I had no answer, but I knew it was time I figured it out. With no plans or prospects, I headed to Sacramento for no other reason than it was the capital of a state I loved. As I had done so many times in my life, I walked into a building and got a job. This time, it was with the California Department of Fish and Game where, thanks to my BS in biology, I was hired as a researcher. Before long, I was being sent around the state to take on various projects. It was on an assignment in Lake Tahoe that I was to discover my destiny at last. I would say it happened out of the clear blue sky, but in fact, it occurred during a storm. A snowstorm kept me from doing my usual work – circling the lake in a speedboat to check on fish that were caught that day. To pass the time, my boss and I hunkered down by the fireplace in our quarters and played a game of chess. Our conversation was punctuated by my grumblings about the ongoing abuse of the environment. My boss stopped playing for a moment and leaned back.

"You know," he said, "you sound like Leopold."

"Who's Leopold?"

He turned to the bookcase, pulled out a book, and handed it to me. It was *A Sand County Almanac* by Aldo Leopold. After reading only a few pages, I was overcome by the connection I felt with the subject and the writer. Leopold was a born generalist who saw the natural world – including human beings – as inextricable parts of a single whole. In my own life, I was losing patience with all the specialists I had encountered in college and on the job. In my view, their narrow focus led to environmental policies that caused more problems than they solved.

Leopold, on the other hand, integrated environmentalism with philosophy, ethics, and pragmatism. And he wrote in a simple, poetic, and accessible style that everyone could relate to.

Although Leopold's work was news to me (like Michelangelo's sculptures were when I visited Rome), he was then – and continues to be – a revered and influential conservationist whose books have sold more than two million copies worldwide. Born in 1887, he attended the Yale School of Forestry. While still a student, he rejected the conventional wisdom of the time that natural predators like wolves and mountain lions should be killed off so deer and cattle could proliferate for the benefit of hunters and ranchers. In fact, he sought to redefine the entire concept of wilderness, from a place that provides human recreation and hunting grounds to a complete ecosystem that supports all of nature.

A founder of the science of wildlife management, his career was full of "firsts." He developed the first comprehensive management plan for the Grand Canyon, wrote the first game and fish handbook for the Forest Service, and proposed the first national wilderness area in the Forest Service system. And in 1933, with his appointment as professor of game management in the Agricultural Economics Department at the University of Wisconsin, he became the first professor of wildlife management. While at the university, Leopold purchased eighty acres in Wisconsin's sand country. The one-time forestland was a barren example of the abuses of environmental mismanagement, having been clear-cut and overgrazed by dairy farmers. As he describes so movingly in *A Sand County Almanac,* he and his family put his theories into practice, planting thousands of pine trees, restoring prairies, and bringing the land back to life. In 1949, shortly after finishing the book, Leopold died of a heart attack while fighting a wildfire in his neighborhood.

Leopold is best known for a seminal idea that he expressed with his usual eloquent simplicity. He urged us all to "think like a mountain," to move beyond our human-centric point of view and see the natural world "as a community to which we belong." The last few words of *A Sand Country Almanac* are still the best description I know of the greatest challenge facing environmentalists – "building receptivity into the still unlovely human mind."

Thanks to Leopold, I had finally found my career, my calling, my drum. I knew it would set the course for the rest of my life. And it has.

I bought a copy of *A Sand Country Almanac*, inscribed it with "my drum," and sent it Sue, the pretty and adventurous guide and translator I had met briefly in Barcelona. We had stayed in touch, and when we were both back in the States, our paths would cross now and then. Once, when we went snow camping together in Colorado, I watched her delight in absolutely every moment of our trip, unfazed by the winter winds and the frozen ground under our sleeping bags. I thought to myself, "You'd better propose to this girl before someone else does. She's one of a kind." It took awhile, but eventually I found the courage and asked her to marry me.

Sue and I were very much in love, and she believed in my dreams and abilities. Still, I'm sure she was relieved to read my inscription and discover that I was beginning to shape a plan for the future because at that moment, I was, at best, a wandering minstrel with little promise of a real career.

As the snow melted and the tourists began to invade Lake Tahoe for the summer, the area was quickly losing its charm for me. One sunny afternoon, as a group of us were having lunch outside in the snow, we heard a commotion coming from the water. A flock of geese had taken off and was heading our way. Small at first, their size increased as they got closer. Soon they were directly above us, heading due north. I took it as a signal and decided to follow them.

A day later, I resigned from the California Fish and Game Department, hitchhiked to the Reno airport, and bought a ticket to Juneau, the capital of Alaska. I found a low-priced hotel, got a room, and started investigating the town. It was a rainy evening. The next morning, it was raining harder. Harder still that afternoon. I read in the local paper that the annual "All Races and Religions Celebration and Picnic" was

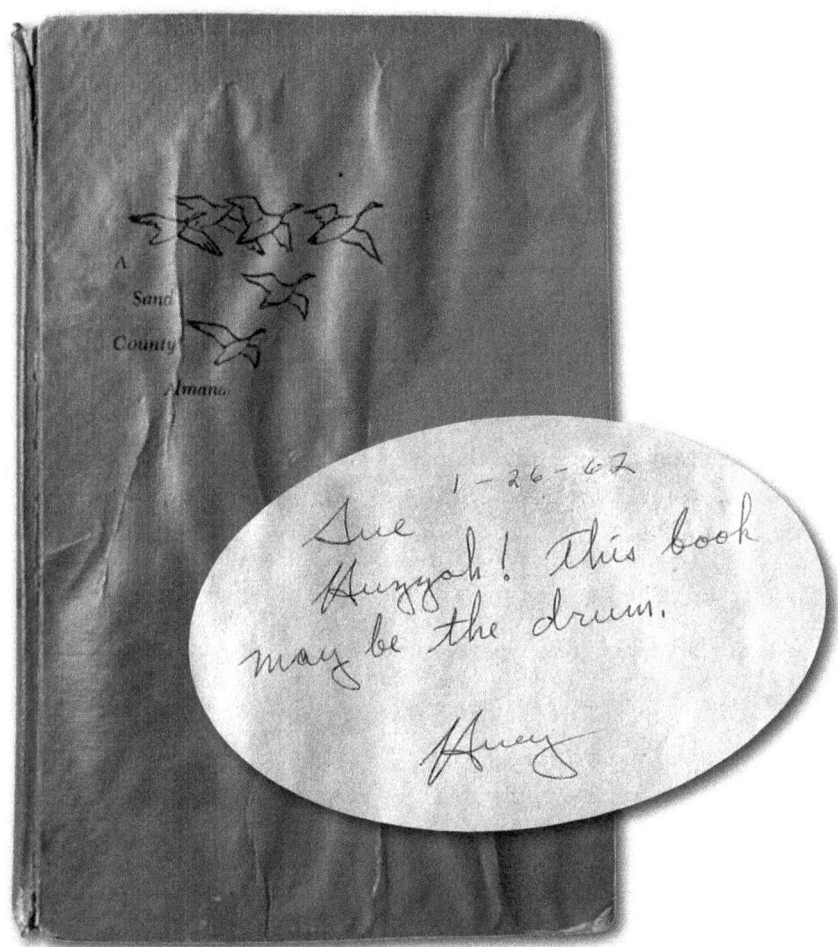

A note to my fiancée after I discovered A Sand County Almanac *by Aldo Leopold.*

scheduled to take place in the local park that afternoon. The desk clerk assured me the event was fun and would be held even in the rain. He loaned me rain gear, and I was off to the festivities. There was a crowd of several hundred people eating picnic lunches, playing softball, relaxing on blankets; all oblivious to the pouring rain. I remember a tray of potato salad floating in rainwater on a picnic table.

Lesson learned: If you're going to live in Alaska, you'd better get used to the rain.

The next morning I went to the headquarters of the Alaska Department of Fish and Game. An hour later, I was a gainfully employed research biologist for the ADFG. I was handed a tag number and sent off to a sporting goods store to stock up on outdoor gear – boots, rainwear, insect repellant, and other stuff. When I returned, someone drove me to a seaplane terminal. Soon I found myself on a floatplane heading to an area near Glacier Bay National Park. Less than forty-eight hours had passed since I'd left my job in Lake Tahoe.

As I looked out of the floatplane window, I caught a glimpse of the wondrous place that was about to be my home. That breathtaking first impression would soon be followed by countless sights, sounds, tastes, and smells that are indelibly etched in my mind and heart: crystalline skies at dawn; sunrise breaking on Mt. Fairweather; regal bald eagles and noisy ravens landing near my feet; killer whales surfacing on either side of my canoe, their whooshing breath mixing with my own; ancient totem poles carefully placed by their creators to look perpetually out to sea; fresh-caught salmon roasting on a fire, its fatty skin cracking in the flames; sweet, juicy cobbler made with wild berries we picked onshore; the pandemonium when a bear got to our food in the middle of the night.

The plane landed on the water off Icy Straight Point. Next to us was a large floating barge with a house on top. In Alaska, these structures are known as wannigans. Ours was outfitted with a large kitchen and bunk rooms containing a number of beds. Several aluminum skiffs with outboard motors were attached to the boom.

Each researcher was assigned to a team that focused on a particular research project. Every morning, I would join my teammates on one of the skiffs, and we would ride out to our project site. When our day's work was done, we rode back to the wannigan and shared a communal

| *Fishing in Alaska.*

dinner. Groceries came by seaplane once a week, but we preferred to pocket our grocery allowance and live off the abundance provided by the land and sea. We set crab traps each morning, and by evening, they were loaded with more crab than we could eat at a meal.

With my lifelong love of fishing and our location smack dab in the middle of the richest salmon habitat in the world, I looked forward to spending every free minute I had in pursuit of that sweet, luscious fish. Yet after just a few outings, I gave it up. For the rest of my stay in Alaska, I never fished for salmon again. Why? For one thing, it was too easy; the salmon practically jumped into the boat. For another, I was already spending all my waking hours in the company of fish – tracking them, smelling them, eating them. So as much as I thought I could never have enough fishing in my life, it took just a few days to find out I was wrong. It was a classic case of getting too close to what you love.

Lesson learned: Dreams and reality usually have very little in common.

My team's project was to find and identify salmon spawning routes. Early on, we tagged the requisite number of salmon as they made their way from the river to the sea. Like all Pacific salmon, these fish would remain in the ocean for seven years before returning to the same river where they had been hatched and undertaking the grueling process of swimming upstream to spawn and die.

During spawning season, our work changed. We were tasked with the job of recovering tags from salmon that had already gone upstream to spawn. It wasn't practical to return to our wannigan every night, so the department hired an old retired fisherman to house two of us on his small boat and transport us along the shores of Admiralty and Chichagof, huge islands on the coast of the Tongass National Forest. Known by the US Forest Service as its "crown jewel," the Tongass's seventeen million acres make it the largest national forest in the United States. According to the Alaska Wilderness League, it is "one of the last remaining intact temperate rainforests in the world." Not surprisingly, it has been a source of political, economic, and environmental conflict for generations. As of 2009, its old growth area is designated as roadless with no logging permitted. Still, the battle continues.

Our assigned recovery area was as large as some states. Each morning the fisherman dropped us off at a new outlet where we would go ashore and follow a spawning stream inland for about half a day. We waded into the water, carrying spears on ten-foot poles and looking out for the bright orange tags attached to dead salmon that had finished spawning or live ones still moving upstream. After retrieving several tags, we walked back to the little boat anchored offshore, slept in our bunk beds, and performed the same work in a different stream the next day.

When the salmon returned from the sea, the wildlife along the streams was magnificent. Every creature imaginable feasted on the heaping tons of dead salmon – brown bears and caribou, wolves and foxes, seals and mink, ducks and crabs. I especially remember the eagles. They were molting during the salmon run and would position themselves on low-hanging tree branches just inches from the water. From there, it was easy for them to bend over and grab the salmon with their beaks.

Our routine was interrupted one day when a commercial fisherman asked our hired fisherman to anchor off a stream mouth in order to block a large salmon run that had entered the stream. Our fisherman explained to us that this was a lousy ploy used by some commercial fishermen to illegally harvest thousands of dollars' worth of salmon. If we had agreed to be their accomplices, they would have proceeded to stretch a net across the mouth of a stream, then go upstream and dump a bag of detergent in the water. That would have driven the salmon downstream in a desperate attempt to get back to the sea, and landed them in the net. To protect spawning fish from these schemers, streams with the largest runs were often monitored by permanent watchmen.

It was wondrous to live in a wilderness area where humans must adapt to animal behavior, not the other way around. Admiralty Island is home to about sixteen hundred brown bears, the largest concentration of their species anywhere on earth – more than inhabit all of the lower forty-eight states combined. Each day as my partner and I made our way up and down the island streams, we were sharing the space with the greatest salmon lovers in the world. If we wanted to stay alive, it was imperative that we understood how bears think and act.

We learned that brown bears are territorial, with a hierarchical society and no known predators. The largest, meanest ones stake out the best fishing spots and protect them fiercely, while the less powerful wander along the stream's edge looking for whatever is left behind. Bears compensate for their poor eyesight with acute senses of smell and hearing. At first sight, nearby bears often took us for other bears, but the downstream wind soon identified us as something else. Not only that, streams often took sharp unexpected turns that could lead to a close and sudden encounter with an enormous and angry bear poised in a threatening stance to ward off competitors.

As we walked, we blew on police whistles held between our lips. Nine times out of ten, a bear who heard the noise would graciously let us by, stepping aside with a slow and noble gait. Still, they didn't always retreat, so if you found yourself looking one in the eye at close range, you had to make some noise and act more threatening than you knew you were.

Soon after the run was underway, several of our crew were attacked by bears because they ignored what we had all been taught. The most dangerous offense was walking on land instead of wading in the water. After gorging themselves on salmon, the bears often went ashore to take a nap. A human that stumbled onto a sleeping bear would have an angry thousand-pound opponent to deal with.

When we were first hired, the ADFG distributed one high-powered rifle to each two-man crew. It was an Alaska magnum, a 300-caliber gun. After a number of bear attacks, each crew member was issued his own. I never shot directly at a bear, but I did fire over the heads of several of them that ignored my whistle and kept coming at me. The sound of the gun turned them around, leaving me intact with a vivid and permanent memory.

Once you've lived in Alaska, it's hard to resist telling bear stories. Here's another one:

We received a radio call from headquarters asking us to substitute for a guard who usually protected the mouth of an important stream. We were to be there for several days. It was deer season at the time. I thought about how a venison steak would make a fine meal on the barge. So I volunteered to ride our skiff ashore and go deer hunting. The stream we were protecting from salmon poachers was at the bottom of a steep walled canyon carved out by millennia of erosion.

In the canyon, I found no sign of deer. To avoid bumping into any fishing bears, I cautiously followed a forested game trail that went up, up, up the canyon wall. When I reached the top, I finally sighted a deer and killed it. Then I had to figure out a way to descend several thousand feet carrying the deer. I took only the hindquarters with me, leaving the front legs, ribs and head behind. I draped the hindquarters across my neck and shoulders and started back down the game trail. I was aware that carrying a freshly killed deer in bear country posed some risks.

Carrying the heavy carcass down the steep hill was trying. At one point, I came to a deep gully that blocked my way. I noticed a huge log that had fallen across the gully, forming a bridge. I started across, but the log turned to powder under my weight, and I crashed into a thicket of devil's club, an aptly named bush full of thorns.

Dragging myself and the deer out of the thicket was exhausting. I stopped to rest and heard a sound that could only have come from a large animal. In high alert, I slowly made my way down the canyon wall, my rifle pointed in front of me and my whistle between my lips. I continued to hear the noise of a large moving animal. The habitat was so thick, I couldn't see what it was, though with my deer meat over my shoulders, I had a good idea. At last, the game trail I was following arrived at the stream and I waded in, much relieved.

Without knowing why, I suddenly felt a sense of foreboding. Terrified, I crouched down and backed away from the bank where I had just waded into the water. I slowly turned to see a bear looking down at me from the banks above a steep part of the stream. He was just a few yards away, staring at me, his head resting on his paws.

I could see that he was close enough to reach out and rake me with his claws. In one motion, I sprang into the center of the stream, blowing my whistle and shooting off my gun, all of which successfully frightened the bear. I saw him do a full back flip and head the other way.

I thrashed my way downstream, towing the deer carcass back to our boat waiting offshore. The venison steaks were delicious.

As thrilling as it was to live in this wild and beautiful place, the isolation and limited number of companions got to me sometimes. It got to everyone. So after weeks on the barge, I was more than excited by the prospect of visiting an inhabited enclave onshore. We rode for several hours on our skiff before arriving at a fueling depot that serviced the large boats moving up and down the strait. Built on stilts like many coastal outposts in Alaska, the depot included a grocery store, a bar, and a loading dock. Civilization!

I steered our skiff into the little harbor. The other crew members jumped off and headed to the bar, leaving me alone to tie up. I found a large post in a safe spot, tied up, and went to join my buddies. We were there for hours, enjoying the change of scenery, picking up some groceries, and having a few drinks.

After a while I noticed that many of the customers at the bar had gone outside. They were lined up along a railing looking down at the pier. I say "down" because in the time since we had arrived, the water level had gone down but our boat hadn't. Without realizing it, I had tied up at a post that was embedded deeply into the mud below the water, not the floating piers that rise and fall with the dramatically changing tides Alaska is famous for. As a result, our boat was hanging in midair, dangling from the post a dozen feet above the water's surface. Because I had left Juneau on a seaplane and lived on floating boats and barges ever since, I'd never noticed the extreme changes in the water level, and no one had ever pointed it out to me.

Everyone howled with laughter as I climbed down a ladder, shinnied up the stationary pole, took out my belt knife, and cut the line. The boat slapped hard on the water. It remained upright, thank goodness.

Living and working on the Alaskan coast.

Opportunities to visit a fuel depot were rare, and we spent most of our evenings on the wannigan playing cribbage. A two-person card game, cribbage is wildly popular in Alaska; locals describe it as the Alaskan national sport. Virtually every boat, bar, coffee shop, and wannigan seemed to have a game going on day and night.

One evening I was challenged to a game by someone I hardly knew. He was staying on our wannigan for a while but wasn't a member of our regular crew. We bet a beer on each game. Over several evenings, I had quite a run of luck, and my opponent owed me seventeen bottles of beer. We weren't allowed to have alcohol on board so the beer was collectible if and when we went to town – which almost never happened.

As my opponent lost game after game, he grew more and more agitated, shouting and cursing with every defeat. All at once, he leaped to his feet, lifted the table with the deck of cards and cribbage board on top, ran out on the deck, and threw everything overboard. Still screaming, he came back below and went for me, intent on a fight. The other guys jumped on him and cooled him off. That strange incident was an example of an Alaska phenomenon – cabin fever. Jack London wrote about a similar, more extreme case when two longtime boatmates grew so sick of each other that they sawed their boat in half, leaving themselves stranded.

Characters like the cribbage player were abundant in Alaska, unusual people who floated from place to place, never finding comfort, never feeling at home with other people. Square pegs, they would quickly become bored and move on to the next spot. Since Alaska was as far as they could go, many chose to disappear into remote and solitary locations where they would work as fishermen, miners, and trappers. And while I liked many of these people, I didn't want to be one of them. So as much as I loved Alaska, and in spite of my being offered one of the rare year-round jobs at the ADFG, I decided to leave the state and start building a more stable life and career for myself and my fiancée.

I headed to Denver because I still had business contacts there from my days at Visking. Making use of my corporate experience, I worked in sales during the day and attended night classes at the University of Colorado. By the end of the school year, I had earned a teaching certificate and gotten a job at an Idaho public school starting the following fall. I wasn't sure whether teaching was for me; still, I thought I'd give it a try.

But what to do for the summer? I'd had my fill of being a salesman again so I took the opportunity to return to Alaska for a few more months. This time I was hired by a commercial fishing company. The money was good and went directly into my pocket since our wannigans and groceries were paid for. I loved being back on the Alaskan coast, but my companions were rougher than the government biologists I had been with the summer before. My new coworkers drank heavily and fought often, and because they were paid by the fishing company only when they delivered their hauls to canneries in town, they ran side businesses selling salmon to people in the tiny fuel depots along the shore. Technically – or not so technically – they were stealing from their employer. Their extra cash led to more drinking and more fighting, and by the end of the summer, I was ready to spend my time with some fresh-faced junior high school students in Idaho, where I had been offered that teaching job.

    I was assigned the position of public speaking teacher at a public school in Boise. I didn't realize it at the time, but Boise, the state capital, proved to be good experience for my future career in government and advocacy. The townspeople – including the parents of my students – were all right-wingers who believed strongly in their right to use public lands however they saw fit. So I set out to have their children look at the world from a broader perspective. Frank Church, Idaho's Democratic senator at the time, had introduced a wilderness protection bill that the local press was strongly opposed to. I asked my students to bring in all the articles and editorials they could find concerning the bill. When they pinned their research on the classroom walls, it became clear that only one side of the story was being presented in the press. That realization disturbed the students and led us to a more balanced discussion of the issues. For a while, some of the parents were demanding that I be fired. But when the students and their parents had the chance to weigh both sides of the debate, most of them became advocates of Senator Church's bill, which did pass the Senate and become federal law. That simple exercise taught me a lesson too: Given all the facts, most people will side with what's best for the environment.

    In my time off, I was a founding member of a search-and-rescue team to help people in the area prepare for emergencies and enemy attack. They were a fun and lively group of volunteers, most of whom I had met mountain climbing. It was the mid-sixties, the height of the Cold War, and the local population was fearful – much too fearful, in my opinion – of a Soviet nuclear attack. The state of Idaho asked our group to plan a drill that would test the preparedness of the National Guard in securing a nearby utility plant from infiltration. Our group's task was to try to get a "bomb" past the guardsmen and into the facility itself. A young woman in our group came up with a creative idea. She would dress quite provocatively to distract the guards from a burlap bag she would be carrying. The bag would have the letters "B-O-M-B" printed across the front. The scheme worked beautifully. As the guardsmen stared at the woman in her scanty outfit, she breezed through the blockade, walked inside the plant, and showed everyone the "bomb" she was carrying. Our team celebrated our triumph at a nearby bar. The guys in the National Guard, furious and embarrassed, did not join us.

It had been more than two years since *A Sand Country Almanac* had inspired me to become an environmentalist. In Idaho, I realized that if I was ever to realize my life dream, I had to make it happen for myself. I am eternally grateful to Utah State University for their open-minded willingness to let me build my own independent curriculum in wildlife biology. I based it on Aldo Leopold's land ethic philosophy, which would later become the foundation of the modern environmental movement. I told my department that I wanted to be a generalist instead of a specialist – an idea at odds with the goals of almost all scientific research. Surprisingly, they agreed.

    I still had to complete certain core courses, which I did without much enthusiasm. For one class, I developed an experiment that tested how crickets adapt to the colors of their surroundings. It went quite well, which thrilled my professors because they could publish the results. I was awarded an MS degree in wildlife biology and began looking for a position doing what I loved.

I don't know when and why science became such a highly specialized pursuit. The great scientific minds of earlier centuries – Galileo, Newton, Darwin – were polymaths and generalists whose genius lay in seeing the connections in nature. Nature *is* interconnected, and the more we work with that reality, the better chance we have of protecting our natural world. Years later, I was vindicated as my wide-ranging generalism became a key asset in navigating the complicated world of California state politics and global environmental advocacy.

That master's degree would never have been mine were it not for Sue, who had generously agreed to marry me and even more generously worked to support my education. Thanks to Sue, I was free to be a full-time student for the first time in my life, reading countless books on a wide variety of environmentally related subjects – forestry, water, grazing, and the burgeoning field of pollution. It was a luxury and a joy.

Sue wasn't the only one who was endlessly patient with my protracted journey to a career. Her lovely parents were too. When we eloped instead of having a proper wedding at their home near Chicago, they accepted it with good humor. And when I continued going to school to pursue a profession they had never heard of, they were agreeable, never pressuring me to become a lawyer or accountant or businessman. If they had reservations about my future, they didn't mention it to me. Sue loved me, and that was enough for them.

Armed with my new master's degree, I applied for jobs at dozens of conservation organizations. I would have been thrilled to be associated with any of them, but I got no offers. Not one. Unwilling to return to corporate sales to earn a living, I applied to a PhD program in wildlife biology at the University of Michigan and was accepted.

I was miserable in graduate school. After about three months in Ann Arbor, I made an appointment with the dean to talk about my deep dissatisfaction with my course of study. When I walked into the lobby of his office, I noticed someone tacking a piece of paper on the department job listings bulletin board. I glanced at the notice and jotted down the information. I hurried through my meeting with the dean, walked directly to the nearest phone booth, and called the number.

The notice was for the job of my dreams in the city I'd always wanted to live in – western regional director of The Nature Conservancy, headquartered in San Francisco. I was determined to make it mine. The person I spoke to on the phone told me that although the job had already been filled, he was willing to meet with me if I could come to his office in Washington, D.C., the next morning. I got myself there.

Somehow I brought to bear everything I'd been reading, thinking, and feeling for years. I convinced the interviewer that I had the knowledge, commitment, and passion to successfully lead the organization's new western division into the future.

More than fifty years later, Utah State University awarded me an honorary doctoral degree. I'm still humbled by the memory of my leading the column of graduates through the auditorium and onto the dais. When I addressed those young graduates that day, I talked about my own youthful struggles and disappointments, how I had learned to believe in myself and trust my own vision of what I wanted my life to be. It was then that I realized if I was going to take the road not traveled, I had to be persistent and not take no for an answer.

As years went by, persistence turned out to be the single most important quality I brought to the environmental movement. All the successful environmentalists I know share this quality. They strive and persevere for as long as it takes – sometimes as long as thirty or forty years – to restore a river, remove a dam, or save a coastline. Perseverance is what makes it happen.

CHAPTER FIVE

Lesson Learned | A Little Craziness Is Good

Saving land is always a story about remarkable people – their personalities, their backgrounds, and their special reasons for defending and preserving open space. During my nine years as western regional director of The Nature Conservancy, we preserved thousands of acres of open land in the western United States. Every one of those acres represents a personal relationship I treasure to this day.

These passionate, driven individuals are often motivated by the prospect of the imminent development of a place they know and love. I was privileged to join forces with many memorable people, enlisting the resources of TNC to realize our mutual goals.

As time went on, our team at TNC grew more sophisticated and skilled at the process of acquiring wildlands. We learned about real estate financing, invented strategies for land gifting, uncovered tax benefits for our donors, and wrote detailed manuals so other TNC staff members could follow our lead. By the seat of our pants and not realizing it at the time, we invented a new profession – strategic public land acquisition.

Now a global organization and major force in the environmental movement, The Nature Conservancy was founded in Virginia in 1915 as a volunteer group called The Ecological Society of America. In the early days, its members fell into two camps: those who thought scientific research should be paramount and those who believed the society should work to protect open spaces. By 1951, when the society became a nonprofit and changed its name to The Nature Conservancy, the conservation advocates had prevailed. Today its mission is "to conserve the lands and waters on which all life depends." Understandably, TNC first focused on acquiring land in the eastern United States where it was founded. In 1960, they made their first purchase in California, but the East remained their main area of emphasis. When I was hired in 1963, TNC was still quite small. In fact, they had only brought on paid staff a few years earlier. I was their eighth employee.

I had no idea what a transformative decade was ahead for me or TNC when I unceremoniously left my University of Michigan PhD program to fulfill my longtime dream of becoming a full-fledged environmentalist. It all started when I saw that job listing on the dean's office bulletin board, rushed to a pay phone, called The Nature Conservancy, and talked the job interviewer into meeting with me the next morning, despite his already having offered the job to someone else. Using all the sales skills I could muster, I somehow convinced him to hire me instead of his original choice. One day, I was a discontented graduate student; the next day I was the western regional director of The Nature Conservancy, responsible for the thirteen states west of the Mississippi.

I went back to Washington a second time for several days of training, then my wife and I drove our cars from Ann Arbor to the Bay Area. It was the heyday of the sixties, and a friend told us about some hippies who had fenced off an area on a Pacific Ocean beach cove where you could camp out for just 50 cents a night. The hippies would even watch your belongings while you were away during the day. We were still pretty broke so we headed there, spending our first few weeks in California living al fresco at lovely Muir Beach, just north of the Golden Gate Bridge. Quite a change from our Midwestern apartment. Looking out at the ocean from Muir Beach, I could see the Spindrift headland to the right and the Green Gulch Ranch to the left. Both areas were privately owned, but before several years had passed, I would acquire them all to be public and preserved forever, two by gift and one by purchase.

When I began my new job with TNC, I was given two directives from the board. One: "Raise money. We're on the verge of bankruptcy." Two: "The local office in Berkeley is currently staffed by a group of grumpy retired University of California botanists. Get rid of them." If I had been more experienced, I would have been intimidated by these challenges. But in my naivete, I wasn't worried at all. I got rid of the botanists by moving the TNC office to San Francisco, then conveniently forgetting to take along the desks used by the Berkeley staff. They got the message.

I rented a small, inexpensive space in a skyscraper. When I arrived the first day, I found my secretary at her new desk, weeping. "What's wrong?" I asked. She said, "You're going to fire me." I said, "Why would I fire you?" She said, "Because I can't type." I said, "You're right. I think we might as well clear the decks." So there I was, the only employee of The Nature Conservancy west of the Mississippi. Being so far from headquarters gave me the freedom to pretty much do whatever I wanted. I went to work.

Shortly after joining TNC, I met the first of the many exceptional individuals I alluded to earlier. His name was Ted Steele, and he often described himself as mentally unbalanced, at least when it came to his obsession with saving land. As far as I'm concerned, he was a great man. His single-minded commitment led to our saving some of the most significant wild space in Arizona, including the Patagonia-Sonoita Creek Oasis in the Sonoran Desert and the Canelo Hills Cienega Preserve. Later, Ted generously introduced me to several other wealthy donors, which resulted in the preservation of the world-renowned Seven Sacred Pools of Maui.

My first meeting with Ted can only be described as bizarre. The national office in Washington, D.C., suggested I hold an annual meeting and deliver a speech to our chapter in Tucson, Arizona. With several hundred members, it was one of the largest in the organization.

I spoke with the local TNC chairperson, the dean of the botany department at the University of Arizona. I arrived with my speech and a slideshow ready to go. Unfortunately, it was raining in torrents that evening, and the only people who showed up were the chairman and his wife. We waited and waited, but no one else came.

Suddenly, a large glass door burst open, letting in a rage of wind and rain. A wild, inhuman scream filled the room, followed by a drenched and pathetic figure. The water poured from his clothes, leaving a puddle under his feet. His hair was matted down in rivulets. With a strange, sad cry, he repeated the same sentence again and again. "They killed my trogon. They killed my trogon."

The dean's kind wife sat him down and wiped the rain from his face. She cooed gently to ease his pain until he was able to shift from screams to quiet sobs. As I watched this weird and mysterious spectacle, the chairman finally explained what was going on. This poor man was actually a prominent bird preservationist in Arizona – Ted Steele.

Once he calmed down, we began to talk. He explained that he had spent the last several weeks in a remote canyon observing a pair of brilliantly colored birds called trogons. Rare in the United State and uncommon in Mexico, trogons feed their young in hollow trees. Each day, Ted returned to the canyon to revel in the birds' company. Then one morning they were gone. Feathers were strewn on the ground, suggesting violence.

He spotted a group of people approaching the area. They stopped and identified themselves as scientists from the Smithsonian Institution in Washington. After chatting with the distraught Mr. Steele, they continued down the path. Ted concluded that these scientists had killed the birds in order to stuff and exhibit them in the museum. I never found out if Ted's suspicions were true.

My strange encounter with rain-soaked Ted was the beginning of one of the most wonderful friendships of my life, and an opportunity to learn a lesson that has guided me for more than fifty years as an environmentalist: There are people with obsessions that nearly drive them insane, but if you can help them harness and channel their passions, they can accomplish great things.

I returned to California. A week later the phone rang. It was Ted. He was breathless and agitated. "You have to get down here immediately."

He explained that a beautiful, rare forest on an oasis in the lower Sonoran Desert had been sold for development and was going to be cut down *the next day*. The forest was a mecca for bird-watching because it attracted birds not found in the surrounding desert landscape where no trees could grow. Evidently, a rare geological phenomenon had pushed an underground stream aboveground for about a mile before it went underground again, creating a rare source of water that allowed trees to survive.

I was on the next plane. Ted had talked to the developer the evening before. He agreed to wait a few hours before starting to bulldoze the area so I could talk to him too. Ted was waiting for me at the airport. We made a quick visit to the site, a place so lush and magnificent, it took my breath away. The developer intended to turn it into a golf course, using the rare water source to irrigate the fairways.

As we were about to enter the developer's office, Ted grabbed me. "You have to save that lovely place. You go ahead inside. I can't stand to talk to the guy." In I went.

I explained to the developer the importance of the oasis as a potential nature preserve. I told him how desperately we wanted to save it.

"Well, I like your appeal," he said. "I like nature too, but I need to make a living. Plus I've already spent money getting this far. The bulldozers are at the site waiting for a signal to start knocking down trees."

I told him I would like to buy the forest and asked him what it would cost. He said a profit over his costs. We negotiated and came to an agreement. We shook hands on the deal, a longstanding western tradition, and I left.

Ted was ecstatic. In a burst of optimism, he had already arranged a celebration with a group of elderly bird-watchers. But there was still one catch. We had six months to raise the money to pay for the oasis or the agreement would expire.

Back in San Francisco, I hired a graphic designer to prepare a fund-raising brochure, and told him to print a supply of ten thousand. No sooner had the boxes arrived and were stacked on the floor of my small office, I got another breathless call from Ted.

"I've arranged a lunch and site visit for a very wealthy woman. I think she'll help. And then I heard the phrase that was becoming increasingly familiar: "You have to come down."

Ted had a job as the greeter at The Arizona Inn, a famously fashionable and expensive residence hotel. When he managed to keep his emotions under control, he exuded charm. The woman he had invited to join us was Cordelia Scaife May, a frequent guest at the hotel. One of the nation's wealthiest and most philanthropic women, "Cordy" was the grand-niece of Andrew Mellon, the founder of Alcoa Aluminum, Gulf Oil, and Union Steel. She loved nature.

I was on the first plane the next morning. The three of us walked through the Patagonia-Sonoita Creek oasis amid the local flycatchers. "How much is it?" she asked. "I'll make a gift to pay for it." And she did.

When I got back to my office in San Francisco, check in hand, I tripped over the boxes of brochures I had hoped would raise enough funds to save the oasis. Now they were useless, thank goodness.

Today the oasis remains as beautiful as ever. The vermillion flycatchers are still there in abundance. They live peacefully in the forest and bathe in the creek that flows through the preserve.

Ted and I were becoming quite a team. Ted would look for land in Arizona that should be saved, and I would "come down right away" and attempt to acquire the property. Ted continued to use his position at the hotel to cultivate relationships with wealthy, generous people who shared his commitment to conservation. (One of them was John D. Rockefeller, who spent his last days as a permanent guest.)

Ted was never one to rest on his laurels. Soon after our success acquiring the oasis, I got a call from him about a new place that needed saving. When I arrived in Tucson, Ted described it as a *ciénega* – the Spanish word for a spring-fed wetland. It was owned by a dying cowboy who wanted it saved.

On the drive out to the *ciénega*, we were joined by two brilliant scholars, Joseph Wood Krutch, a retired Columbia University professor, National Book Award winner, and nature columnist for the *New York Times*; and Paul Martin, a geoscientist and paleobotanist at the University of Arizona. Joseph, having moved to Tucson for his health, became an avid student of the southwestern environment and wrote many influential books on desert ecology. Like me, he was deeply affected by the teachings of Aldo Leopold. Paul was well-known for his theory that prehistoric humans of the Pleistocene era had developed hunting tools that led to the worldwide extinction of many large mammals.

The four of us arrived at an ancient adobe house with an old pickup truck parked in the yard. We approached a figure leaning on the truck.

"Howdy, boys. Here's my place. I'd appreciate it if you would evaluate it to see if it's worth preserving."

On the hood of the truck was a half-empty half-gallon of red wine. "I'd like to go with you to the *ciénega*, but, as you know, I'm awfully sick. This is my painkiller," he said, patting the wine bottle.

We took a nearby path down to a deep canyon not far away. There was a magical cottonwood forest at the bottom and a live stream flowing vigorously through it. At the head of the canyon was another stream of water gushing out of the rock wall. Schools of minnows darted about. The earth was blanketed with blooms of a small wildflower that stopped us all in our tracks.

"What flower is that?" I asked.

Paul said he didn't know. He had never seen it before. Joe said the same. Only later did research reveal that this lovely wildflower was a rare Mexican species, and that this *ciénega* was the northernmost place it had ever been found.

We walked back and reported to the cowboy that we were awestruck by the magnificence of the canyon and would be honored to save it for him.

"One thing I ask," he said, "is that you carve out a building site on each of the four corners of the property. I would like to leave one to each of my children."

Years later, I mentioned the *ciénega* in passing to the singer Linda Ronstadt, who had grown up in Arizona. As we chatted further, she stopped me and began to describe the *ciénega* herself. She owned the ranch next door.

Yet another of Ted's discoveries was the Wood Brothers ranch at the mouth of Aravaipa Creek near Phoenix. Ted had ferreted out the spot and found that it blocked access to a steep walled canyon where the river flowed about knee-deep. Its banks were steep, unroaded, wild, and glorious. We negotiated with the owners and got their agreement to sell us Aravaipa Creek. As always, Ted energetically pursued funding from his network of donors and got it paid for. He even went to federal officials and convinced them to designate a huge adjacent area as wilderness, expanding the area many times over.

What drove Ted to such extremes to save wild places? It was a question I asked myself often, and one night in front of his huge stone fireplace, I found out. Originally from Kansas, Ted had moved to Europe to escape his parents' disapproval of his being gay. At first, his family supported him financially, but eventually he found a profession that he excelled at and that earned him a good living – card sharking. Working the bridge tables on ocean liners that sailed between the United States and Europe, he won high-stakes games against wealthy travelers.

One day a woman he knew slightly from previous crossings was playing bridge at his table. She asked him to join her for dinner where she revealed the real reason for her invitation. Marriage.

"Ted, I know you cheat at cards, still you're charming and I like you. I live in Tucson, I'm very wealthy, I live alone in a large fancy house, and frankly, I'm lonely. I like evening social events very much, especially in the winter when many of my friends from the East are vacationing in Tucson. Going alone is uncomfortable. I have all the money we would ever need, so you could stop this silly business of preying on elderly travelers. There must be some dangers in doing that. So Ted, I'd be honored if you would marry me."

I can see Ted now, dressed in a blue sports coat and cravat, smiling and responding sincerely. "My dear, I like you very much, but marriage would never work. You see, I'm gay."

"Oh, Ted, I don't care about that anymore. My husband's been dead for some years. What I want is a companion, and you are just so likable."

Before long, they were married.

And that is how Ted became a land saver in Tucson, and why we were having drinks that evening in the large stone mansion where he lived with his delightful wife. They had extended a permanent invitation to me to stay with them whenever I was in Tucson, which I often did.

Ted's story wasn't over yet. He asked that a bottle of whiskey and a glass of ice be brought over to the fireplace. While I had a beer, he proceeded to drink the entire contents of the bottle.

"Huey, I'm driven by an awful reality. I killed my mother." He looked at the glowing embers. After a while, he went on.

"She was suffering from a serious cancer and was in great pain. She begged the doctors for more painkiller because it wasn't doing much good." She pleaded with Ted to end her life and free her from her constant pain. It wasn't a thing to do, but Ted did it.

"I will never get over it, Huey. I live with the burden every day."

He added sadly, "So I'm driven to save life." Which he did for the rest of his own life.

As an environmentalist and a human being, I learned lessons and acquired skills from my relationship with Ted that have helped me throughout my life.

Ted and Mrs. May taught me the importance of maintaining close ties with donors. The trust and friendship I established early on with a small group of devoted benefactors have endured to this day. In fact, I was able to found a new organization in 1972, The Trust for Public Land, thanks almost entirely to the generosity of Mrs. May. TPL has gone on to preserve more than three million acres of open space nationwide.

I also learned from Ted and his wife not to view property owners as adversaries. Instead, I always try to show them that I'm reasonable and understand their point of view. I ask them straight out what they would need to make a deal work and what barriers might interfere with an agreement. For example, the developer who sold us the land at the Patagonia-Sonoita Creek oasis was happy to cooperate once he found out that we were willing to make him whole financially. The cowboy at the Canelo Hills *ciénega* had more emotional concerns – a legacy for his children. I also assume that land owners have at least some love of nature. After all, many of them have lived their lives in exquisite, unspoiled settings.

Yes, there are times when fighting is right and necessary, but it's surprising how often I've been able to win through negotiation, reasonableness, and empathy.

Ted Steele is also at the heart of what I consider the most important wilderness I ever saved during my fifteen years in professional land conservation. It encompasses a swath of land in Maui from the rim of eight-thousand-foot Haleakala National Park Volcano down to the ocean on the windward side of the island. It includes a steep and lush jungle canyon and the utterly stunning Seven Sacred Pools near Hana. The project is also the best example of how Ted and I practiced the new profession of strategic public land acquisition that we had developed in Arizona.

As usual, it all began with a call from Ted. This time, instead of saying, "You have to get down here immediately," he said, "You have to go there immediately," meaning Maui. His bird-watching friend, Frances Baldwin Cameron, was a descendant of one of the original missionary families of Maui. He talked to her once in a while, and on this occasion, she was upset about a pond and wildlife refuge that had been in her family for generations. During World War II, the area was loaned to the US military. They never returned it. Over the years, the family tried to get back the property they legally owned, but the military found ways to stall and put obstacles in their way. The sudden urgency for wresting control of the property was prompted by the military's recent decision to fill in the pond. The Kahului Airport runway was now adjacent to the refuge, and the government was concerned that the migratory birds who wintered in the pond would collide with an airplane and cause a crash.

Ted continued: "Get over there. She practically owns the island. Her family own a newspaper. They own the sugar business. I told her about you. She wants to talk to you about her problem." And in typical Ted fashion, he added, "Don't call. Go over. Now."

So I went. I talked to Frances and figured out a plan. First, I hired some college students to sit at the end of the runway and count the birds as they flew in and out of the existing pond. I also learned that there had never been a recorded instance of a bird smashing into a plane at this airport. Armed with evidence, we conveniently made use of Frances's powerful daily newspaper. In articles and editorials, we described in detail the colorful history of the pond and the rare species that called it home. Once the royal fishpond of King Kamehameha, it was a migratory habitat for birds from Asia, Alaska, and the mainland. There were even pintail ducks from California who wintered in the pond.

What made this challenge so unusual was that instead of our wanting to buy a private tract and make it public, we wanted to acquire (ostensibly) government land and return it to its original private owner. Still, our goal was the same as always – preserving open space.

My PR strategy created some heat, even in Congress, but soon the word came from Washington that the land would be returned to its real owners. Today it is a National Natural Landmark known as the Kanaha Pond State Wildlife Sanctuary. A 143-acre wetland, it is home to two endangered Hawaiian bird species, the Hawaiian coot and the Hawaiian stilt. Any visitor leaving the airport should pull over and watch the birds in the pond. It is a beautiful experience.

While talking to the Camerons on one of my visits, the conversation turned to one of the elder members of their family who had just passed away in Hana. "His home was in such a glorious, remarkable spot, and now it's got to be sold." Apparently, his children had never considered the taxes they would owe upon his death. "There's no way they can afford to keep it."

About a month later, I was in Hawaii again working with Mrs. Cameron on the bird sanctuary. Having finished up what we needed to do, I flew back to San Francisco. I had just landed at the airport when I heard myself paged on the loudspeaker. For once, it wasn't Ted Steele. Instead, it was the president of The Nature Conservancy calling from Washington. "Laurance Rockefeller wants to have lunch with you in the worst way in Hawaii. Please get on the plane and go back to Hawaii." That's all I was told. So I went home, changed my clothes, said hello and good-bye to my family, and headed back to Hawaii. My boss was referring to Laurance Rockefeller, a third-generation member of the Rockefeller family and one of the greatest conservationists in American history. He was presented the Congressional Gold Medal in 1991 for having contributed to the protection and expansion of hundreds of open lands, including national parks in California, Hawaii, Maine, Vermont, the Virgin Islands, and Wyoming.

Things got awfully complicated from there. I arrived in Honolulu expecting to get a connecting flight to Maui and meet with Laurance straightaway. Instead, someone I didn't know met my plane and told me I was going to be attending a meeting at the airport. I entered a large meeting room where lots of people were seated around a huge table, all yelling at one another. Their heavy wool suits told me that they had just flown in too, probably from New York.

"There's the guy that cheated Laurance!" screamed one of them.

Defending myself, I hollered back, "What are you talking about? I don't even know why I'm here."

Another chimed in, "Oh, shut up, Charlie, he didn't know anything about it. Our attorney asked him here to help."

I listened for a while and picked up on the drift of their problem. It had started several weeks earlier at a cocktail party where Laurance was chatting with a charming young woman he knew casually. A transplant from Georgia to Hawaii, she worked as a real estate agent.

Evidently their conversation went something like this:

"Laurance, remember that beautiful area I told you about? It's up for sale. I've got the listing and its slated for development."

Laurance was crushed. "Oh, my dear, that would be terrible."

"I know! Can't you do something?"

"I'll call the bank in the morning."

Which he did, without checking out any of particulars at all. Only later did he and his attorneys learn that the agent had misrepresented the property. Her listing was for only one part of what she had claimed at the party. When Laurance Rockefeller's staff found out what had happened, they were furious and immediately flew to Hawaii to deal with the mess. That's when I came on the scene.

When I finally found out what and where the property was, I understood why Laurance had acted so fast to buy it. It was the now world-famous Pools of 'Ohe'o (Seven Sacred Pools), a place of unique beauty and historic importance to all Hawaiians. Sculpted by the volcanic flow from an ancient eruption that wended its way from mountain to sea, it formed a series of little streams and waterfalls that splashed and danced in spectacular fashion.

At this point, Laurance Rockefeller owned two of the pools, although he had been led to believe he had purchased all seven. Only later did I connect the dots and realize that the Seven Sacred Pools were part of the ranch that the Cameron family had been forced to sell in order to pay their taxes.

We protected Maui's Pools of 'Ohe'o (Seven Sacred Pools) from imminent development.

Within a few seconds of my arrival at the airport hotel meeting, the man in charge of Rockefeller real estate resigned and walked out. After a moment of shock, the group went back to arguing. Seeing no way I could help, I walked outside and sat on a park bench where it occurred to me that five of the Sacred Pools might still be on the market. I concluded that the best use of my time would be to fly immediately to Maui, spend some time studying the place in question, and get a handle on the situation. I still hadn't been contacted by Laurance or his staff, but I wanted to be informed and prepared when he was ready to talk.

After I arrived in Maui, I made the long drive to Hana, looked around the area for a while, and then checked into a nearby hotel. The handsome bar turned out to be a goldmine of information. The bartender, a native Hawaiian woman named Daisy Nelson, listened to my tale. "What you're doing is important to us. We don't want a hotel there. We want to save the pools. I think I can help." As the evening went on, a succession of people came up to me at the bar, expressing their hopes of saving the pools. With every conversation, I became more committed to finding a way to keep the Seven Sacred Pools wild forever.

In the meantime, Daisy had called a local Realtor named Hamilton McCaughey and asked him to meet me at the bar. Handsome and gregarious, he was the subject of much tongue wagging among the Hawaiian social elite. According to the gossip, Ham had befriended a wealthy Texas couple who were having an extended stay in Hawaii to resolve some problems in their marriage. Supposedly, he lured the wife away from the husband. Soon after they were divorced, she and Ham were married. The fact that she was a Waggoner, one of the richest ranch families in Texas, only added to the gossip. By the time I met Ham, he and his new wife were persona non grata in Hawaiian society, although I found him a pleasant and charming character.

Far more important to me – and the reason Daisy had invited Ham to the bar to meet me – was the fact that he and his wife had bought the rest of the Cameron Ranch. So they were now the legal owners of the five pools Laurance Rockefeller mistakenly thought he had purchased. We talked for a long time. He related a sad story about his late father who had been a professor at the University of California at Berkeley. Driven to contribute more to the world than he thought he was doing at the university, his dad changed professions and became a fund-raiser for the Audubon Society. Evidently, he wasn't suited to the job, and having failed in his new career, jumped to his death from the Golden Gate Bridge. I sensed that the son, like Ted Steele, wanted to do something positive in his parent's memory. The next day, after talking it over with his wife, he offered to sell the property to us so we could save it. Then and there, Ham became another one of the special individuals who made possible the preservation of a precious natural space.

Once he and his wife agreed to sell, we began negotiating the price. As always, I was grateful for my sales experience as a young man. I suspect Ham had also been a salesman at some point because he was a skilled negotiator too. We met in a big, fancy house trailer he had installed on his newly acquired land. It was situated at the edge of a steep cliff overlooking the lush Kipahulu Valley. As we started going back and forth, we were interrupted by a delightful man who stopped by for an unexpected visit. I'm convinced his arrival helped lighten the mood and ultimately eased the path of our negotiations.

Tall and thin, he reminded me of Ichabod Crane, and he accepted the nickname with good humor. His real name was Bob Wenkham, and as we got to talking over a beer, I learned that he was a prominent author and photographer who made gorgeous photo books about wild Hawaii. Ham invited Bob to stay for dinner. Eating our burgers, we looked out over the canyon and asked Bob if he had ever done a book on Maui. He said no. "Well, why don't you?" "Okay, I will." And he did. The book, *Maui, the Last Hawaiian Place,* is a treasure not only for Bob's spectacular photos and language but also for its two forewords: one written by one of my heroes, David Brower; the other by Charles Lindbergh, who would soon prove instrumental in acquiring the Seven Sacred Pools.

After our delightful interlude with Bob, Ham and I returned to our negotiations. First he drafted an agreement to his liking. Then I did the same. Then we tore up the agreements. Next, I went for a walk and he wrote up another one. Then he took the walk and I tried another version. Eventually we got to yes. Ham and his wife sold the property for a million dollars, accepting my need to get credit and time to raise the money. The three of us had dinner to celebrate.

I still had heard nothing from Laurance Rockefeller about the meeting I was summoned to Hawaii to attend. So I sent word to the Rockefeller attorneys that I had just negotiated an agreement to buy the rest of the property, and I flew back to San Francisco. When I arrived at the San Francisco airport, it was, as Yogi Berra used to say, déjà vu all over again. I heard my name over the loudspeaker. I flew back to Hawaii.

This time, Laurance and I did meet each other, at one of his hotels. "My gosh, how did you do that deal? Let's go have lunch up on the mountain and celebrate." As we walked up the long hill to some waterfalls, I looked behind me and saw a procession of hotel staff following us with preparations for our lunch – tables, chairs, umbrellas, coolers full of food. They looked like ants climbing an anthill. "Man, I wish just once they'd leave me alone," Laurance said. He would have been happier with a sandwich.

We arrived at the edge of a forested canyon so lush, it was literally a green wall. I was told there was no record of humans ever having ventured inside. We sat at a table. "Congratulations," he said. "Now, how are you going to pay for it?"

A good question.

As we talked, I thought about the dense jungle and steep canyon in front of us. It gets four hundred inches of rain at sea level and rises eight thousand feet to the volcanic rim. I thought some more. I knew that twenty-eight species of birds – all lovely little honeycreepers – were listed as extinct on Maui. If there was anywhere in the world where a bird thought to be extinct might be rediscovered, it would be somewhere in the secret places hidden in this canyon.

Why did I care so much about extinct honeycreepers? Because I knew that bird-watchers were – and continue to be – the largest and most passionate group of nature lovers in the country. And it didn't hurt that many of the board members of TNC were avid bird-watchers too. Some were authors of ornithological books and manuals. I was convinced that if we could galvanize the interest of bird-watchers, we would raise the million dollars to buy the land in Hana.

I rolled the dice and said, "Mr. Rockefeller, The Nature Conservancy will do it."

"Do what?"

"We'll save it."

"Really?"

"Yes."

"Well, fair enough if you want to try."

"Yeah, we'll do it."

We had lunch and shook hands and walked down the mountain.

What I didn't say that afternoon was that the TNC board was very conservative and had never negotiated anything this big and expensive. Frankly, I committed to Laurance Rockefeller before getting the board's approval because I knew they would have said no.

So there I was – in for a penny, in for a pound, I faced a burden of more than a million dollars. Plus I still had to tell the home office the news. I wasn't sure which challenge would be harder to pull off.

Back at TNC headquarters, I stared out at the unhappy board and presented my case, arguing for the grandeur of the project and the value of doing a favor for Laurance Rockefeller. As I had hoped, they concluded they had no choice. Along with their reluctant approval, they made sure I knew that it was up to me to raise all the money, and that I would be fired if I didn't.

I laid out my strategy. Step One would be rediscovering an extinct bird. I called Richard Warner, a young biologist I knew at UC Berkeley. "Come on over to Hawaii. You're going on an expedition." He was eager. Joining him was Marty Griffin, the physician and founder of Audubon Canyon Ranch who had worked with me to save a lovely island in the Bolinas Lagoon, as well as a few other California scientists. Then I hired a couple of Hawaiian cowboys to help. I went to Safeway in Honolulu and bought all kinds of food in waterproof boxes for their journey to one of the wettest places on earth. I watched the group head into the green jungle wall with machetes flashing. They had to climb eight thousand feet in soggy, muddy terrain, so I knew it would be a week or two until they returned. I waited.

When the expeditioners emerged from the wilderness, they had indeed rediscovered the honeycreepers thought to be extinct. Whew! I took the film Marty had shot, put together a slide show, and produced a little postcard for fund-raising purposes.

After completing Step One, I started working on Step Two, planning fund-raising parties where I would show my slideshow and hopefully raise a million. The first event was set for Honolulu. I was lucky enough to line up Charles Lindbergh, the pioneering aviator and a Hawaii resident, to give a speech. All of Honolulu's rich, elderly, and charitable citizens showed up.

A problem came up I hadn't anticipated. When Ham and his wife arrived, it caused an uproar because of their reputation. Even worse, Ham joined the speakers circle, which was a special honor. That was a mistake. The other guests were visibly upset, and the focus of the evening changed from saving the Seven Sacred Pools to the offense of the couple's presence.

To make matters worse, the appearance of Charles Lindbergh caused an uproar. During his heyday, he was adored by teenage girls and young women, the very ones who were now the elderly dowagers at our fund-raiser. Suddenly, these women were teenagers again, mobbing Lindbergh, pulling on his clothes and grabbing his body from every direction. He pushed the women away without success. Then he picked up the microphone in front of him and hurled it across the room where it bounced off the far wall.

Things finally calmed down. I somehow managed to present my slideshow, and Lindbergh gave a fine talk about the beauty and grandeur of the Kipahulu Valley. But the evening had already been ruined. Of the million dollars we had hoped to raise that evening, we got zero. Not one cent.

The first party may have been a disaster, yet I had no choice but to carry on to the next scheduled event in San Francisco. The host was George Jewett, heir to a large timber fortune. He contributed money himself and agreed to host the party at the Saint Francis Yacht Club on the waterfront. We sent invitations to the city's elite. When only a few responded, we "papered the walls," a term used in fund-raising for offering free tickets to whomever you can so you don't wind up with an empty room. We called a number of environmental groups. The hardworking nonprofit employees and volunteers who attended weren't used to such lavish occasions – or the complimentary gourmet treats they were served. They planted themselves next to the buffet tables and followed the servers who were carrying trays of food. Mr. Jewett came up to me and said, "The manager says that this crowd just set a record for eating oysters."

Once again, we raised almost no money, about $10,000 all told. Only $990,000 to go. I started to get worried.

Chicago was next. Using our Rockefeller connection, we got the Wrigley family to host the affair with the help of a professional event company. The owners of Wrigley chewing gum and one-time owners of the Chicago Cubs, the family brought together a good crowd. We finally raised some real money, but considering our costs, it was nowhere near enough.

New York was the last stop on our party tour. Everything was riding on it being a success. We hired a planner and sent out invitations to

Laurance Rockefeller's private list. An invitation from Laurance amounted to a command request. It gathered a lot of acceptances.

On the day of the event, the fund-raising began at a small lunch – so small, in fact, that the only attendees were me, Laurance, his wife, and two other people. One was Doris Duke, the North Carolina tobacco heiress. The other was a member of the Wallace family who founded *Reader's Digest*. Before lunch was served, a waiter placed a small bowl of water on the table in front of each person. I didn't know what the heck to do with it. Seated on my right, Mary Rockefeller could tell I was a real country boy. She leaned over and said, "That's a finger bowl. You dip your fingers in it."

After lunch, I stood up, showed the group my photos, and gave them a fiery pitch. Seated next to Laurance, Doris Duke patted his arm and said, "Laurance, I think this is a lovely idea. I'll give a quarter of a million dollars." Then the other woman patted Laurance's arm and said, I'll give a quarter of a million too." The main event hadn't even started, and we were halfway home. I was feeling pretty good.

In spite of his ordeal at the party in Hawaii, Charles Lindbergh was kind enough to attend our fund-raiser in New York. Earlier in the day, I worked with him on his foreword to Bob Wenkham's book – the very same book that was thought up over beer and hamburgers in Ham's Hana house trailer. It was quite an experience getting to know Lindbergh as a person. After making his historic ocean flight in 1927, he became perhaps the greatest American hero of all time and certainly the biggest celebrity of his era. When I met him, he had long ago retreated from public life, and I appreciated his coming forward on our behalf.

The hour arrived, and the cocktail party began. Several people spoke, including Lindbergh. I got up and gave my slideshow, then set out to seal the deal. "We're asking for money to save the Kipahulu Valley, the rare birds, and the waterfalls." Somebody stood up and said, "I'll give twenty-five thousand." Then someone else: "I'll give fifty thousand. Then another and another. We exceeded a million dollars so fast that it was all over in minutes. I'd never been in a situation like that before, and I didn't know what to do. I finally stopped the appeal at about a million and a half. Along with the money pledged at lunch, the day was a great success. With our project funded and my job secure, I slept well that night. I was especially happy that no one ever found out that Laurance had been taken in by a Realtor in Honolulu.

Looking back at it all, I realize that there was quite a bit of luck involved in the process. What if the ornithologist and the cowboys hadn't discovered the extinct birds in the canyon? What if I hadn't met Ham in the Hana hotel bar? What if the New York cocktail party had been a disaster like the ones in Honolulu and San Francisco? But it all did happen, and we were able to protect one of the most beautiful wild places on earth. And despite the skepticism of The Nature Conservancy board of directors, they wound up with their first million-dollar acquisition – not to mention long-lasting relationships with some of the wealthiest philanthropists in America.

When the news arrived at TNC's Washington office, the board sent me a letter of congratulations. Buoyed by success, I had an idea. What if we could add to the land we had just acquired by annexing the canyon and valley nearby? Then we would have the entire area from the volcano to the sea. It's all state-owned anyway. So I went right to Honolulu, and got hold of the Rockefeller attorney whose problem I had solved by acquiring the Hana property. I explained that I wanted to get the state of Hawaii to donate several thousand acres of Kipahulu Valley, which would establish a significant national treasure. After the terrible experience with Laurence and the Realtor, he realized how vulnerable these precious areas can be. He agreed to help. He gave me the name of a respected lobbyist whom I hired on the spot. I wasn't sure if hiring a lobbyist would be kosher with TNC so I "forgot" to mention it.

Our strategy was to reach someone in a high position with state government and enlist his or her support. We discovered that the speaker of the assembly, Elmer Cravalho, had been the first mayor of Maui before becoming the first speaker of the Hawaii House of Representatives after statehood. We got an appointment with him. The speaker walked in like royalty. He was short and climbed on a box behind the table so he could look down on me. "What is this about?" I took out my maps and told my story. I had outlined the area we were interested in. He said simply, "I'll support it on one condition. I get to announce it." I said, "Fine." So he had a press conference and announced that the valley would be donated to the National Park Service to preserve Hawaiian history and heritage. I find it ironic that this far larger area turned out to be much easier to acquire than the original land – and we got it for free. That's the way it is with preservation. It's hard to get the ball rolling, but once you do, well-meaning people will step up to help.

Lesson learned: Adding on is always easier than getting started.

So that's how the Haleakala Crater got connected to the sea. But there's still one more piece to the story. When I went to Hawaii to celebrate the opening of the park, I took a taxi from the airport to the edge of the crater. I had brought along a backpack and a sleeping bag with a plan to sleep at the rim, wake up at dawn, and hike twenty miles into the crater and out the other side, all in one day. It was raining and kind of cold. As I hiked along a splendid trail, I grew tired and decided to take a nap or meditate for a while. I sat down near a huge and astonishing mahogany tree so I could admire and remember it. I fell asleep. When I woke up, I looked at the tree, and I swear it was dancing. Although I am not particularly superstitious, I thought, "My God, something's going on here." I ended the long day at the hotel bar in Hana where everything began. I said hello to Daisy, the delightful Hawaiian bartender who had told me she would try to help me save the pools and did. "Daisy, something happened today that I don't understand.

Then I asked her, "You know where that mahogany tree is?"

She said, "Yeah, I do."

"I took a nap next to it, and when I woke up, I swear it was dancing."

She said, "Oh, yeah."

"What do you mean, 'Oh, yeah?'"

She said, "That was Lindbergh thanking you."

She told me that Lindbergh had died recently and was buried very near the mahogany tree. I remembered a similar experience I'd had in New Zealand, also at a crater. The Maori held the spot in great reverence, and they would hold their hands over their eyes when they passed a certain sacred place.

That's all I have to say about the dancing tree, but for some reason, it has always felt like a fitting end to my magical time in Maui.

Ted Steele tops my list as a human being and a land saver, but there were many other caring souls who made my stint at The Nature Conservancy so productive and memorable. Like Ted, Hamilton McCaughey, and that Arizona rancher with his truck and wine bottle, they all had a story and a reason for their generosity.

Charles Borden was my first "customer" at TNC. When I had finished moving into my new San Francisco office, I sat down to go through a stack of mail that had piled up during the move. The first letter I opened was handwritten and so beautifully composed, I thought some of my friends back in graduate school had sent it to me as a practical joke. But the return address was local, Muir Beach, where my wife and I had camped when we first arrived in California. The letter asked TNC to save a piece of land in Marin County. It was signed

"Charles and Eleanor Borden." I was excited and called my wife right away. "Why don't we go out and look at it because I don't believe it's real." So we drove there and located the spot, Spindrift Point, a hilly headland jutting out into the sea beside Muir Beach. The property was protected by a barbed-wire gate and a strange-looking fence. When I rattled the fence, a guy walked over to us, opened the gate partway, and looked us over with suspicion. He was short, had a bright seaman look, and wore a dark knit watch cap pulled low over his forehead. After introductions, he invited Sue and me inside the property and showed us around the spread, which appeared to be about a couple of acres in size. The man turned out to be Charles Borden himself. It was the first of many visits I would make to Spindrift over the years. Charles and I became lifelong friends.

By the time I met Charles, he was in his fifties and had settled down from a life of great adventure. He had gone to sea at thirteen and spent decades working on freighters and exploring the islands of the Pacific on his seventeen-foot boat. Having circumnavigated the globe four times in his tiny vessel, he was a legend to single-handed sailors (people who sail long distances alone in small boats). An excellent writer, he authored many books and articles about his travels, the most famous being *Sea Quest: Global Blue-Water Adventuring in a Small Craft.*

When he and Eleanor decided to finally put down roots, they selected the extraordinary spot at Spindrift and hired prominent architect Henry Hill to design a home that would resemble a ship. Charles also had his builder carve out a small cave-like writing studio from a rock at the western end of the property. It had space for two chairs, a small desk, a potbellied stove, and, on the floor, a bunk like you would find on a boat. The bunk, which looked directly down on the Pacific, was a re-creation of his quarters on the rusty old freighters where, in his down time, he had begun his writing career. Every afternoon around four, he would walk a narrow path along the ridge of his property to his refuge in the rock. A night owl, he wrote until midnight, then read into the early morning, snuggled up against his window above the sea.

As I got to know Charles better, he often invited me to join him in the cave. The two of us spent many days on those two chairs, talking and philosophizing about the future of his land and the future of the world. The most memorable times were when there was a storm blowing in off the raging sea.

Sometimes he would call me in the late morning and ask me to join him in North Beach for lunch – or, in his case, breakfast. Whenever Charles left Spindrift and came in to the city for lunch, he reminded me of a sailor on shore leave. I identified with that feeling because I had spent some time as a ship's mate myself. While at sea, I'd obsess about something like an ice cream cone and couldn't wait to get ashore and have one. Charles was the same way. He came "ashore" by heading to a little Italian restaurant that a friend of his owned in North Beach, San Francisco's old Italian neighborhood. We were always warmly welcomed and seated at the same table.

Between visits to his home and lunches in San Francisco, I guess we met at least forty times to talk over what would happen to his land after he and Eleanor were gone.

Charles introduced me to literature in a way I had never experienced before. He said that while working on ships, he would spend every free minute he had reading. He even acquired the skill of reading while standing on a deck that was pitching to and fro. When I would show up at Spindrift, I'd invariably find him with his nose in a book. He would look up, his eyes squinting, and blurt out with great passion, "God, have you ever read so and so?" Usually I hadn't and said so. Once I found him reading *Lord Jim*, which he handed to me. "Have you ever read Conrad?" Predictably, I hadn't. "Conrad was Polish. Can you imagine a Pole learning English and writing it perfectly?" His passion was infectious, and I wound up reading countless books that he recommended.

He and Eleanor became regulars at my family's holiday dinners. My children loved them. After he died, Eleanor continued to join our dinners where she would retell Charles's sea tales. They were great stories and made us feel like he was still around.

For instance, she recounted a story about a young friend of Charles who died on a ship while they were in Hong Kong. Charles wired his parents who wired back, "Please bring him home." With no refrigeration on board, Charles came up with his own plan for preserving the body. He got a container, put his friend's body inside, and filled it up with cheap liquor. It worked and he delivered the body to the parents intact.

Charles knew his way around booze, having run liquor on ships during Prohibition. He and his pals put bottles of vodka in the ship's deck hold, one layer deep, then covered them with soft plaster. On top of the plaster, they poured on black gunky oil. When the inspectors came aboard in their white uniforms and white gloves, they took one look at the sticky, oily mess and left. The ship sailed on to San Francisco where guys in the know came aboard, went directly to the hold, broke up the plaster, lifted out the vodka, and packed it up for delivery to speakeasies across the city. That's how Charles got the money to buy Spindrift Point. It's probably the only preserve in The Nature Conservancy's history that was paid for by rum-running during Prohibition.

Another story starts with Charles sailing on Monterey Bay. He saw a woman on the shore painting at an easel. She yelled to him, "Boy, aren't you afraid you're going to sink in that little boat?" They started talking and a few days later, they were married! Charles and his bride loaded up his boat and took off on their honeymoon. They were gone so long that an obituary was placed in the local newspaper. In fact, they did have a harrowing trip. Eventually they arrived in Hawaii, but they were not in good shape. Neither of them could stand up, and they had to be carried to shore. As soon as they recovered, they got divorced.

Charles believed – or rationalized – that he needed to have love affairs to make his writing passionate and vibrant. He often had several extramarital relationships going on at once and composed rapturous love letters to each woman. Always the writer, he kept copies of every letter in a big, fat file in his desk. I think he really did see the letters as a literary exercise because he would often show me the latest one and ask me to critique it. When I heard he had died suddenly of a heart attack, I immediately called his doctor, Herman Schwartz, who was also my friend. Herman and I both raced to Charles's studio, hoping to get to his desk and remove the file before his wife found it, but when we arrived, she was already there, love letters in hand. She was quite calm. "Hi, guys. I know what you're doing. I knew all along, that rascal."

He was a man of such passion. Even the way he died exemplifies his devotion to his land – he suffered a heart attack chasing trespassers from his beloved Spindrift Point. I'm happy to say we bent some rules and buried Charles on his own land overlooking the sea he knew and loved so well.

It was my great good fortune that he was the first person I met when I entered the land-saving business. Personally, he was my friend, teacher, and inspiration. Professionally, he set the standard for hundreds of people and projects to come. Principled, energetic, and obsessed with seeing Spindrift Point protected forever, he also made sure that Eleanor was allowed to live there for the rest of her life. I have always felt an obligation to have Spindrift carry on as he and Eleanor dreamed it would. I'm gratified The Nature Conservancy has done what I promised him I would do.

Even if he hadn't generously donated the magnificent piece of land he owned, George Wheelwright would still be a man for the ages. Born into a wealthy Massachusetts family, George spent his youth studying fine arts at Harvard, working as a deckhand on a United Fruit Company ship, and owning a successful summer camp in Santa Barbara. Brilliant and restless, he jumped in and out of new ventures with abandon. After growing bored with his summer camp, he returned to Harvard, this time as a graduate student and instructor in physics and astronomy. In one of the classes he taught, he met a precocious young undergrad named Edwin Land. Land had left Harvard after his freshman year and invented filters capable of polarizing light. Returning to Harvard a few years later, Edwin asked George to look over an idea he had and give him his honest opinion of his work. The next morning, George told Edwin he thought they should start a business together based on the invention. In 1932, the two young men formed Land Wheelwright Laboratories; within five years, it had become the Polaroid Corporation.

In another five years, George was ready to leave Polaroid and Edwin – especially Edwin. His move was precipitated by an encounter he had with Edwin over the Christmas holidays. Edwin lived and breathed work, never closing his lab for weekends and holidays. Good-hearted George, knowing the beleaguered staff would be at work on Christmas Day, stopped by with champagne for everyone. To George's surprise, only Edwin was at work. He was lying on the floor in his lab, pounding on his equipment and crying in frustration about something that wasn't working out right. George, a well-rounded person with many interests and joys in life, took one look at Edwin and quit the company.

It was 1942. George went to Washington and worked with the government on how to apply polarization to the war effort. Then he convinced the US Navy to let him join the military in Europe – even though he was about to turn forty. He trained as a pilot and flew captured German scientists to the United States.

After the war, George ran into Hope Richardson, an old neighbor from Massachusetts. The two were soon married, and, not surprisingly, for George at least, they decided to start an entirely new life as ranchers in California. They bought a rundown cattle ranch in a spectacular location in Marin County near Muir Beach and Spindrift Point. George and his wife spent millions bringing Green Gulch back to life, restoring its home, and improving the quality of the herd. They raised Hope's children there.

George was a progressive and altruistic man. I got to know him and Hope because they donated generously to The Nature Conservancy over many years. His philanthropic spirit affected his children's inheritance dramatically. He left them only a small part of his fortune – just enough for each of them to buy a modest home. George strongly believed his children would live more fulfilling lives if they earned their own way in the world. And he believed his money would be more useful helping worthwhile causes. He was right on both counts.

One day, I received a sad call from George explaining that Hope had terminal cancer. They wanted to talk to me about the future of the ranch. I went out there four or five times, sat by Hope's bed, and listened to them describe their wishes for Green Gulch. Their request was simple and direct: keep the property as it is forever and make sure it will be used for a noncommercial purpose.

TNC's job was to find a buyer who met the Wheelwrights' criteria. First I approached Marin County, suggesting to them that the ranch would be an ideal educational center for Marin citizens. Everyone I spoke with loved the idea except the county board of supervisors. Its conservative wing dismissed the proposal as a waste of money. That was the end of that. I wasn't sure what to do next, but I happened to meet someone at a cocktail party who turned out to be Richard Baker, the abbot of the San

Francisco Zen Center. About that time, I was grousing about the supervisors to everyone who would listen. When I mentioned it to Richard, he said, "Boy, I'd love to own that ranch." Recalling my time in Southeast Asia and how moved I had been by my exposure to Buddhism, Richard's thinking started to make sense.

Presenting the idea to George and Hope, I explained that Zen Buddhists had a reverence for nature. Plus Buddhists were excellent gardeners! They agreed to sell their ranch to the Zen Center for half its market value. Since then, Green Gulch has become a destination for meditation and contemplation, a place where anyone can visit and learn about the Zen way. It has an organic farm and garden, conference facilities, and family programs. Most important, it exists in harmony with its natural environment, and, every time I visit there, I'm convinced it fulfills the Wheelwrights' wishes for their home.

George's life had a fitting and karmic conclusion. He lived out his last years at Green Gulch where the Buddhists cared for him with love and gratitude.

Not far from where George Wheelwright lived at Green Gulch lies the grave of the British master gardener and horticulturist Alan Chadwick, a brilliant and highly influential figure from the counterculture of the sixties. His name isn't well known today, but he played a key role in many of the movements the Bay Area is now famous for – organic farming, biodynamic gardening, composting, and the rise of California cuisine. Alan was a colorful and charismatic character who transformed a four-acre hillside on the University of California, Santa Cruz campus into a lush garden that showed off his revolutionary practices. With the support of a devoted band of UCSC students, Alan and his garden became the most famous attraction on the newly established campus. I've never understood exactly what Alan's official role was at the university – if he had one at all – but one day he was fired. That's when Paul Lee, a professor at UCSC and a mutual friend, called me to ask if I could help find somewhere for Alan to live and continue his important work.

I connected Alan with Covelo Ranch where he lived for a while, then I introduced him to my friends at Green Gulch. My wife had done some volunteer gardening at the center, and we thought Alan might fit in well there. He established himself at Green Gulch, teaching residents and visitors his gardening techniques and practices. Before long, the name Green Gulch Farms became synonymous with local, seasonal, organic produce sold at farmers' markets, natural foods groceries, and especially at Greens, the renowned vegetarian restaurant owned for many years by the San Francisco Zen Center. Today, the eight-plus-acre garden and orchard is the focus of ongoing volunteer programs, apprenticeships, and hands-on learning.

Then there was Charles Watson, one of the greatest minds of the environmental movement. After a long career at the Department of the Interior, he started the Nevada Outdoor Recreation Association (NORA) in 1958. I met him in the late sixties when he showed up outside my office door and sat there for several days without ever asking to speak to anyone. Finally one evening, I was the last to leave and asked him what he wanted. He smiled and asked if he could walk with me. In a short time, he had explained that he was working to safeguard a little known national treasure, the millions of acres of public lands. He was personally visiting, photographing, and mapping unique, beautiful sites. Ultimately, his work resulted in a large book encompassing all his research. It led to many new places receiving protected status by the government.

Charlie had a tactic of enlisting the help of retired government experts who knew from experience how their former departments operated.

One Environmental Protection Agency retiree helped successfully sue the EPA for various infractions. Someone else on the NORA team was a recent retiree from the US Bureau of Land Management. While still at the BLM, he had been under constant pressure to perform illegal political favors. When he began working with Charlie, his inside knowledge helped NORA win a lawsuit with far-reaching implications. The ruling required the BLM to perform environmental impact studies before issuing grazing permits. Their success was huge. As I recall, it increased protection for more than a hundred and sixty million acres.

Of all Charlie's enormous accomplishments, I think his most important was writing the Federal Land Policy and Management Act (FLPMA). It was a sweeping and brilliantly conceived law that transformed a rat's nest of existing laws, rules, and regulations into a comprehensive blueprint for federal land. Almost fifty years later, the blueprint still works beautifully. Charlie defined FLPMA's master plan as the "management of the public lands and their various resource values so that they are utilized in the combination that will best meet the present and future needs of the American people." It also gives the BLM the right to designate and oversee wilderness areas and wilderness study areas. Today, nearly nine million acres of wilderness are under FLPMA's jurisdiction.

Charlie's courage is legend, and he enjoyed playing David to the Goliath of powerful timber, energy, and cattle interests. He was famous for showing up at the conferences of western cattlemen's associations, where he positioned himself in the front row and politely offered comments on the legality – or illegality – of their proposals.

NORA's annual meetings were humble affairs, usually taking place in the back of a café. But along with its members and volunteers, the attendees always included the highest-ranking federal officials responsible for public lands. They understood Charlie's power and influence and knew the importance of keeping him on their radar.

Charlie and NORA were one and the same. His prodigious skills and achievements were the result of his intellect, eccentricity, and a 24/7 dedication to the land. No days off. No vacations. He lived on a meager retirement income, traveling great distances in an old VW bus to attend regional meetings of one sort or another. Once I suggested that he back off his schedule a little, spend some time socializing, maybe get married. He blushed and kicked the ground. "Oh, you know me, Huey. I'm married to the public lands."

Of course, not everyone obsessed with safeguarding wild lands is "a little crazy" like Ted Steele or Charlie Watson. For those people, it doesn't hurt to offer a financial incentive in addition to appealing to their better angels. With the help of a savvy lawyer and great friend named Putnam Livermore, I developed a strategy that sweetened the deal for about a hundred well-meaning donors.

I met Putnam through his mother, Caroline. She phoned me out of the blue when I first started working for TNC. Caroline was the descendant of Robert Livermore, a British immigrant who started ranching in the East Bay in the 1840s, when California was still part of Mexico. Mount Livermore on Angel Island in the San Francisco Bay and the East Bay city of Livermore are named for the family.

Caroline wanted me to help save the Seaman's Church in Tiburon. I immediately let her know that TNC saved land, not buildings. Although I was polite and apologetic, she was not happy with my answer and hung up in a huff. Before long, I received a call from the TNC national headquarters. They had obviously heard from Caroline.

"Don't you understand politics at all?"

I learned that it wasn't a smart move to turn my back on Caroline Livermore, the grande dame of environmental causes in the Bay Area. I was instructed to get busy and help her save the church. I knew what I had to do, yet there was something about the situation that rubbed me the wrong way. It *wasn't* my job, or the TNC's, to spend our time or resources on preserving buildings. I thought about it for a while and came up with a way to resolve my dissonance and preserve the church for the right reason.

I contacted a geologist friend of mine who, unbelievably, had received his PhD in rocks of the Tiburon Peninsula. The two of us found a flower under the church steps that was part of a rare plant called the black jewel. Mrs. Livermore got her church, and I still had my integrity.

But what I really got out of the experience was the opportunity to meet Putnam. He was his mother's attorney and began donating his valuable time to advising me at TNC. Putnam was the one who first explained to me that taxpayers could benefit greatly by donating their assets to charities like TNC. Here's how it worked, and still does:

Let's say a wealthy land owner expresses an interest in donating a piece of property. According to the IRS, a donation of privately owned land to a qualified conservation organization fulfills the requirement for a charitable gift. So the land's appraised value can be taken as a tax deduction, reducing the donor's income and lowering the taxes they owe.

Land saving was one of them. For some reason, TNC had never pointed out this advantage to their donors and potential donors before. Once we made it part of our story, our donations increased significantly. And I will be eternally grateful to Putnam Livermore for making it happen.

After my instructive experience with Putnam's mother, I knew better than to ignore strangers who showed up out of the blue. So when a young hippie-looking guy walked into my San Francisco office with no appointment, I offered him a seat and listened to his story. Good thing I did. That twenty-year-old was Bill Milton, one of seven siblings who had just inherited the thirty-five-thousand-acre Bear Tooth Ranch outside of

Helena, Montana. They wanted to donate the property for a wildlife refuge. I told them I would do whatever I could to help them realize their dream. It's hard to visualize just how big a thirty-five-thousand-acre property is, but to give you an idea, it's about fifty-five square miles, or more than seven times the size of San Francisco.

Acquiring the land would have been a simple, straightforward transaction if it hadn't been for the boys' uncle who was executor of the estate. He was an obstinate, belligerent man completely opposed to the wishes of his brother's heirs. But I met with him on the off chance he would accept my offer to purchase the land. Surprisingly, he agreed and added, "Look, this is an old-fashioned western deal here. Let's just shake hands, and it'll be yours for the price you're proposing." The final step was to present our sealed bid at a court hearing. Shortly before the hearing, I found out that the uncle had entered into a second "old-fashioned western deal" with a rancher who presented a higher bid. The court would be obligated to accept the larger amount.

On the verge of losing the ranch, I did some investigating and learned that the other bidder was the owner of the Redfield Gun Sight Company in Denver and an ardent environmentalist. I called him up, and he suggested we get together in New Orleans where he was attending a conference. We met in a bar where I explained that the heirs wanted their ranch to become a wildlife refuge. He said, "Well, I don't really need the whole ranch. I'm just interested in the eight-thousand-acre hay farm that's part of it. I've got my own ranch up there, but I don't have enough winter feed." I told him The Nature Conservancy would be happy with the remaining twenty-seven thousand acres. He and I agreed on a deal and came up with a plan together: I would bid on the whole package, then turn around and sell him the hayfield he wanted.

The court date arrived. There was a lot of tension in the room. A few days earlier, I had accused the executor of malfeasance, something

I probably shouldn't have done. He was furious and made a scene in the courtroom, threatening to sue me and The Nature Conservancy, and demanding the hearing be postponed. At first, the judge was restrained, but after a while, he lost his patience and let the executor have it. He ordered the hearing to proceed.

The judge reviewed the bid Mr. Redfield and I had secretly collaborated on in New Orleans. The uncle spoke up, not knowing that Mr. Redfield had joined me in the first bid: "There's another bid, Your Honor." The judge corrected him. "No, only one bid has been presented to the court." Which is how twenty-seven thousand acres of Bear Tooth Ranch became a wildlife refuge.

Sometime later, the governor of Montana invited Bill and me to take a ride in a National Guard helicopter for a bird's-eye view of the land. We admired the bighorn sheep, the elk, and the deer. It was our aerial victory lap. The story of the Milton family and their extraordinary donation appeared in the July 4, 1970, issue of *Life Magazine* in a cover article about Americans taking action to protect our wildlands. Coincidently, the article also described how Ted Steele and I had acquired Arizona's Aravaipa Canyon Wilderness for TNC.

Whenever someone says to me, "But, Huey, what can I do? I'm only one person," I get pretty steamed. Passionate individuals like Ted, Laurance, Hamilton, George, Bill, and the two Charlies – and others I'll discuss later, like Wangari Maathai – have protected unique natural places for birds, animals, and plants, and for millions of people to experience with joy and wonder. The lesson they teach us – and we must always remember – is the power of one. And when it comes to saving land, a little craziness is always good.

CHAPTER SIX

Lesson Learned | Never Give Up. Never Give In.

My work with people like Ted Steele, Laurance Rockefeller, and Hamilton McCaughey shows the value of cooperation and empathy in convincing property owners to sell their land for preservation.

But the time inevitably comes when an environmentalist has to play hardball. I have my own term for the process – "creative conflict" – because it takes all the imagination, self-confidence, and toughness you can muster to take on entrenched interests and win. It also takes a talent for ignoring what everyone else is repeatedly telling you: you're fighting a lost cause.

Creative conflict – sometimes called three-dimensional chess – is an effective environmental tool because it leverages existing reality to bring about desired change. The first step in the process is to take a clear-eyed look at the big picture and identify your opponent's vulnerabilities. A person may seem protected by institutions, laws, regulations, and rank, yet he or she is still a human being with emotions, sensitivities, and shortcomings. In fact, the trappings of power often lead to overconfidence and carelessness.

Take politicians, for example. Many of them succeed by keeping a low profile. They make deals and pass legislation under the radar. This makes them vulnerable to public exposure. Bringing their activities to light makes them squirm and can often change their mind. Banks operate the same way, so disclosing their loans and transactions can have a similar effect. Another potent tool is cultivating friends in the "enemy camp." If your cause is moral and serves the public good, there are usually people in the organizations that oppose you who are uncomfortable with the position they're required to espouse. They may be looking for an opportunity to give you valuable information or even cross over to support your cause.

If you've never heard of a place called Marincello, creative conflict is the reason why. In 1965, when America's romance with suburban living was at its peak, Marincello was well on its way to becoming a "new city" in the Marin Headlands, the serene coastal hills immediately north of the Golden Gate Bridge. The plans for the former two-thousand-acre ranch originally included enough homes, malls, high-rise apartments, hotels, and theaters to support one hundred and fifty thousand residents, not to mention a "Venetian canal" that would bisect the shopping center. Over time, the project was scaled back to homes and apartment towers for thirty thousand.

Things moved quickly. Permits were approved, roads bulldozed, construction sites surveyed and staked, and an official "Marincello" gate erected and celebrated at a grand groundbreaking ceremony. The project's Connecticut developer, who specialized in creating complete, self-contained cities from scratch, described the site as "probably the most beautiful location in the United States for a new community." Gulf Oil, now part of Chevron, provided the financing. Like many giant corporations in the sixties, Gulf had begun diversifying into industries like transportation, chemicals, and land and real estate development.

I was already familiar with the property because it was next to a ranch owned by one of our most generous donors, Ann Witter, the daughter of Dean Witter, the founder of the San Francisco–based brokerage firm.

The front gate was the only thing built at Marincello.

So when I got a call from Mel Wax, a first-rate rabble rouser, I wasn't surprised. As the mayor of Sausalito, Mel had successfully led the charge against building a freeway along the town's waterfront. (He was also the anchor of KQED-TV's newsroom, and later, Mayor George Moscone's press secretary.)

"I'm hosting a small meeting Sunday, and I would appreciate your coming. We want to fight Marincello." Based on my TNC experience, I was selected to negotiate directly with Gulf. I made about eight trips to Gulf headquarters in Pittsburgh, meeting with their real estate division on behalf of The Nature Conservancy. Each time I entered the office, I came face-to-face with a gigantic artist's rendering of Marincello in all its glory, rampant with skyscrapers and parking lots. Later, they made a large architect's scale model for the on-site sales office. The Park Service in Marin County still has the model. They haul it out every once in a while for an exhibit.

At first, the Gulf executives wouldn't budge. Why would they? They had everything going for them – ownership of the land, approval of the Marin County Board of Supervisors, participation of the largest and most successful "new city" developer in the country. But I didn't budge

A 3D model of Marincello, on display in the development sales office. Photo by Kenji Yamamoto.

either. Over the course of our meetings, I told them again and again, "It's not going to happen." They smugly pointed out that there was no way our side could afford the legal fees to fight them in court. When I responded that our lawyers were working for free, it was the first time I saw fear in their eyes.

In the meantime, we mounted a multipronged opposition campaign. Our first line of attack was a ballot initiative, yet even that was stacked against us. The county ballot guy threw out a bunch of the signatures, claiming they were improper. They weren't, but he disqualified enough of them that we were removed from the ballot.

We persevered. Three brilliant lawyers – Doug Ferguson, Marty Rosen, and Bob Praetzel – volunteered their time and slowly started to turn the Marincello battleship around. Bob contributed over a thousand hours to the case, which went all the way to the California Supreme Court. The ruling went our way by one vote. Evidently, Gulf had neglected to file some legal documents during the requisite ten-day period. At the time the papers were originally due, the board of supervisors reassured Gulf it wasn't a problem to miss the deadline. But to the California Supreme Court, it was a very big problem. By using creative conflict, Bob had turned a tiny vulnerability into our first victory. We were no longer a lost cause.

At the same time, we started collecting signatures for a new ballot initiative. Two of our volunteers were sisters in their eighties. Their tactic was to walk right up to morning commuters who were stopped in traffic as they merged onto the Golden Gate Bridge. One sister would approach the driver for a signature while the other hit up the passenger for money. They were incredibly successful. I think the element of surprise had a lot to do with it.

Then there was the credit card protest. We urged Gulf credit card customers to cut up their cards and mail them back to corporate headquarters in Pittsburgh. Thousands of them did.

Between the Supreme Court case, the Golden Gate Bridge sisters, and the credit card drive, we were generating a lot of media attention. That, in turn, put pressure on the Marin County Board of Supervisors and the Gulf Oil executives. Exactly what we had intended.

By the time I showed up for my next meeting in Pittsburgh, Marincello had shrunk a little. The vice president of real estate was seated next to a pile of torn-up Gulf credit cards. "What the hell is this all about?" the vice president asked. "Now I've got the oil division mad at us. You're wrecking any chance you might have had for us to be cooperative." I said, "Well, it's obviously a signal that people are not going to let this be developed."

At every meeting, there was this same mystery man seated at the table. He never introduced himself or said a word, but it was clear from his facial expressions that he was not happy with how the aggressive, bombastic real estate VP was handling the meetings and dealing with me. I guessed that the silent executive had been sent from the main part of the company, the oil division, to observe our discussions. I also sensed that he thought Gulf should never have moved into the real estate business to begin with. Playing my hunch, I worked at making the real estate man look ineffective to his superior.

I guess I read it right because at the next meeting, there was a new vice president looking back at me. "Joe's no longer with us." The president showed me one more model of Marincello, even more scaled back than the one before. I said, "No. I don't know how many times I have to say this. The public is not going to accept any development there." Then he shouted, "All right, everybody leave the room." I got up to leave. "Not you, Johnson." I sank back down.

"All right, you son of a bitch, you win. What do I do?"

I was stunned. I never expected to hear those words from an executive of one of the world's largest corporations. I had no words and was just able to sputter, "Win?"

"Yes, he said. "You win. We're not going to develop the site. We're willing to sell you this thing. What do we do now?"

I started to regain a little composure. "Uh, you've got to sign an option."

"All right, I'm here to sign an option. Where's the option?"

Not expecting any of this to be happening, I took out a sheet of notebook paper and wrote out a rough version of an option. The real estate division president signed the notebook paper. Then I remembered that the option wouldn't be valid unless it was accompanied by a down payment. The president asked how much I was putting down.

I said, "A hundred dollars," which was all I had in my checking account.

"What the…"

By this point, the adrenaline had kicked in, and I was thinking clearly. "It's a token, a gesture of good faith. What you're buying is our ability to sell the land to the National Park Service and deliver half the land value to you in cash and half in tax credits against your corporate profits. We can do this for you. You can't do it for yourself. So you're getting an excellent deal."

"All right. Jesus, get out of here."

We went to Congress and got the money. And that was the last I ever heard from or about Gulf Oil.

In March of 2020, Debra Schwartz, my longtime friend and founder of Tam Hiking Tours, generously organized a discussion about Marincello at the Mill Valley Public Library. Sponsored by the Mill Valley Historical Society, it was called Marincello Revisited. Joining me on the panel were Marty Griffin, Bob Praetzel, and Doug Ferguson, all of them well into their eighties and nineties – Marty had just turned 100! – and still fighting for the environment. It had been nearly fifty years since our victory over Gulf Oil, and I was humbled to see the room packed with people too young to remember Marincello. Way back then, we had fought to save the headlands for posterity, and now posterity was thanking us for what we had done. What a wonderful moment it was.

John Hart, Marin County poet and environmental writer described the Marin Headlands of today in his 2003 article for *Bay Nature* magazine:

> Today the Marin Headlands is one of the most visited, most appreciated sections of the GGNRA [Golden Gate National Recreation Area], itself the most heavily visited national park in the country. Four million people, park planners estimate, head over Conzelman Road, through the Rodeo tunnel, or out Tennessee Valley every year, while the land remains prime habitat for coyotes and bobcats and, reportedly, mountain lions. Bird watchers flock to see the resident peregrine falcons and

hundreds of migrating raptors. Hikers, bikers, joggers, and equestrians travel the ridges, in fog or in sunshine. Amateur botanists kneel to the flowers that, in spring, seem to cross the landscape in waves. History buffs look out from the old gun emplacements toward the horizon over which enemy fleets never came. Kids and adults at The Headlands Institute learn of the landscape and its stories. Artists of all kinds converge at the Headlands Center for the Arts. Across the way at the Marine Mammal Center, crews tend to injured seals and sea lions. Nobody seems to complain about the wind.

There's one more critical lesson to learn from the Marincello victory – a landmark acquisition ignites people's imagination and empowers them to go after more land. The Marin Headlands became the anchor property for what was to become the Golden Gate National Recreation Area. The first noncontiguous National Park in US history, the GGNRA's eighty-two thousand acres reach across the coastal areas of San Mateo, San Francisco, and Marin Counties. The park now includes Alcatraz, the Presidio, Fort Mason, Fort Funston, Crissy Field, Sweeny Ridge, and much more. The story of how the Marincello victory led to creation of the GGNRA and Point Reyes National Seashore is well told in the 2012 documentary film, *Rebels with a Cause*.

Saving the coastline near Bolinas involved both carrots and sticks – and a guy I met by chance in a hot springs pool. I had long dreamed of acquiring the pole farms the RCA corporation owned along the ocean from Bolinas to Point Reyes, comprising about three-and-a-half miles. RCA used to grow trees there, then turn them into telephone poles that enabled ship-to-shore radio communication. Over time, technology made the poles obsolete, so I thought RCA might want to sell the land. I wasn't the only one who had that idea. Developers were circling around the area as well. There was talk about building homes, resorts, and golf courses in the unspoiled area. Unfortunately, I had no connections to RCA, and I had learned from experience that without connections to the decision makers, nothing happens.

I still can't believe what happened next. I was soaking in the hot springs at the Tassajara Buddhist retreat near Carmel Valley, talking with a random group of strangers in the dark night. I heard someone speak out, "Does anyone have any use for an artist and a real estate specialist from New York City?"

So thinking about RCA, I said, "Yeah, maybe. Do you know anybody with Radio Corporation of America?"

The voice said, "I have this fishing buddy who's their lead attorney.

I said, "Well, I think we can be friends here." So we talked a while and did indeed become fast friends. His name was Herb Arnold.

Not long after our soak, Herb called me from New York. "I'm going fishing in the morning with my friend from RCA. Can you get here?"

I grabbed a bag, flew to New York, and joined the two of them the next morning. When we got out to sea, I made my case, explaining the tax advantages of gifting the property to TNC. It made sense to him, so he arranged for us to meet the president of RCA a few weeks later. Back in San Francisco, I gathered together the documents for my big presentation. Then I started thinking, "I'll be meeting with the president of RCA, for God's sake." So I borrowed an attorney friend's elegant hand-tooled leather presentation binder for my papers, and my lawyer, Doug Ferguson, and I headed to New York.

The meeting was to take place after lunch at Rockefeller Center, where RCA had its headquarters. Arriving a little early to rehearse our pitch, Doug and I had lunch at an Italian restaurant in the basement. I said, "Okay, you say this and I'm going to say that and I'll hand him this paper."

Ready to go, we were ushered into a conference room with a big round conference table. The door opened and in trooped a line of gray-suited executives who took their seats around the table. We introduced ourselves, and I said somewhat pompously, "Here is our proposal." When I opened my handsome portfolio, we all noticed that the documents were covered with a big pile of spaghetti. Everyone understood that we had just eaten at the Italian restaurant in the basement, and they all burst out laughing. Mortified at first, I soon realized that the spaghetti had actually broken the ice and humanized the situation. RCA agreed to give us their three and a half miles along the coast, which became part of the Point Reyes National Seashore.

A peaceful marshy wetland in the Bolinas Lagoon is yet another example of the power of creative conflict. Named Kent Island, its idyllic location and proximity to San Francisco made it a prime spot for development. Way back in the twenties, some speculators constructed a harbor that was intended to provide access to the racetrack they were planning to build. The racetrack never happened, but the harbor remained. Forty-some years later, the regulars at Smiley's Schooner Saloon in Bolinas came up with another idea: to create a resort complex complete with hotel, restaurant, office space, a fifteen-hundred-slip marina, and even a helipad. They hired a developer and leased a hundred acres of underwater tidal land to be dredged and deposited on the island so it would sit above water.

What encouraged the crowd at Smiley's to move ahead with their plan was the fact that ownership of the old harbor district would give them the right to condemn any private property that got in their way. A widow – the namesake of Kent Island – owned about eighty acres of the site and refused to sell. The developers threatened to condemn the property if she didn't go along. On top of that, the Marin County Board of Supervisors was ready to approve the project. Everyone, including the Marin monthly, *The Pacific Sun*, agreed the case was closed.

I drove out to the island. Seeing it again and experiencing how pretty it was, I got mad as hell. Mad at myself for not paying closer attention to what was going on, and mad at the cynical businessmen who were willing to intimidate a widow and ruin a natural treasure to get what they wanted. I thought about the other times I had taken on a lost cause and managed to win – Hawaii, Marincello, Arizona. I decided to try and turn this one around too.

First, I contacted some people – Marty Griffin, for one. Marty was a young doctor who was passionate about birds and had been part of the Maui expedition to find the long-lost honeycreepers. The two of us went to see Mrs. Kent and explained that we didn't have any money at the moment, but if she would sell her land to The Nature Conservancy now, we would somehow raise the money and pay her the $80,000 the property was worth. She agreed.

Then I went to Peter Behr, a committed environmentalist and member of the Marin County Board of Supervisors. He was a wonderful political leader and had become the golden boy of the board. Peter jumped right in to help.

"Okay, here's what we do," explained Peter. "The supervisors are currently four to three in favor of the project, but there's one guy named Kettenhoffen – he's the chair – who is hated by the environmentalists. It really bothers him. His feelings are hurt. One time, he asked me why everyone was against him. I told him it was simple – he always signs on for every development." Playing to Kettenhoffen's need for approval, Peter went to him. Ket, I'm going to do you a favor. I want you to vote for a surprise gift to the county. It's a park." Ket asked where it was. Peter was cool. "You know that island in the lagoon out there in Bolinas? It has a pending development, but it would be nice if it was left alone."

Peter continued. "I know some people who are willing to acquire it and offer it to the county as a park, but you've got to vote yes in the supervisor's meeting because the developers are going to be mad as wet hens when they see what you did." Ket said, "Okay."

Now it was my turn to act fast. I hurried to get the legal documents drawn up for Mrs. Kent to sign, which she did. Then we arranged with Peter Behr to gift the land to the county that same day, only one day before the supervisors were scheduled to meet. First thing the next morning, I was at the title insurance company, waiting for them to unlock their front door. When they did, I went in and registered the title under The Nature Conservancy. From there, I drove to the board of supervisors meeting, walked in, and standing at the back of the room, I held up the title document.

Peter said, "Ket, we need to interrupt." Kettenhoffen jumped to his feet and repeated, "We need to interrupt. We have been offered a wonderful new park." Everyone expected Ket to vote with the developers as he always had. Instead he said, "I would like to make a motion that we accept it." Ket was finally an environmental hero, and we were glad to have him take the credit.

I'm not sure the supervisors understood what was happening, but they voted four to three in favor of the county park.

The developers who were in the audience jumped up and screamed, "What the hell is going on?" But the battle was over, and they had lost. What they failed to understand was that while the harbor district could condemn Mrs. Kent's property, they could not overrule Marin County, a municipal entity. It was a joy to watch all the yelling and screaming in the chamber, knowing there was no way to undo what had just been done. Thanks to Marty Griffin's fund-raising skills, we raised Mrs. Kent's money very quickly.

A San Francisco Chronicle *editorial cartoon about the fight to save Bolinas Lagoon.*

Kent Island has grown in importance and protection since its designation as a county park in 1967. In 1981, it was made part of the Gulf of the Farallones National Marine Sanctuary, and in 1998, Bolinas Lagoon, including the island, was named one of seventeen Wetlands of International Importance in the United States by UNESCO. Kent Island is also a wildlife sanctuary for many bird and mammal species, among them harbor seals, brown pelicans, and the endangered snowy plover.

We won another "lost cause" in San Diego, acquiring a beautiful wild canyon despite intense pressure from the wealthy, well-connected neighbors in Rancho Santa Fe. Years earlier, a couple who revered nature had bought the canyon next to their home to protect its wildlife from habitat loss and development. They loved hearing birdsong in the morning and coyote howls at night. The residents of Rancho Santa Fe felt differently. They wanted to clear out the canyon vegetation and wildlife they considered a danger and a nuisance. There were unsubstantiated rumors that coyotes had killed children in the area. They had no evidence, of course, but were determined to have their way.

I received a call from the longtime owner of the canyon who told me a heartbreaking story. As the Rancho Santa Fe owners and attorneys intensified their pressure on him to sell the land, his wife became more and more distraught until she reached the breaking point and killed herself. I put my group of brilliant volunteer lawyers to work, and we won the day. What a wonderful legacy for a woman who connected so deeply with nature.

It was at the 1972 United Nations Conference on the Human Environment in Stockholm, Sweden, that I pushed the concept of creative conflict farther than I ever have, before or since.

I had long dreamed of establishing a UN environmental program where people like me could join forces to take on global environmental problems. Obviously, environmental problems are, by their nature, global, and many of us were growing frustrated with piecemeal solutions restrained by local, state, and national boundaries.

A British noblewoman, Barbara Ward, the Baroness Jackson of Lodsworth, used her position and influence to get the idea off the ground. Barbara was much more than just an aristocrat. A well-respected economist and advocate for the developing world, she turned her focus to environmental sustainability in the sixties. Two of her books, *Spaceship Earth* and *Only One Earth* (written with René Dubos originally as a report for the Stockholm conference) set the stage for the way we see our planet today.

With 115 nations involved in four years of preparation, the Stockholm conference was held as the first step in the formation of a permanent UN agency. Its purpose was to devise a plan and structure for the proposed agency that would be presented to and hopefully adopted by the UN General Assembly at a later time.

I arrived in Stockholm with high hopes. I walked into a beautiful castle that was to serve as the conference headquarters. It spanned an entire city block. When I stepped up to the registration table, I was told that I wasn't enrolled and was prohibited from attending any of the scheduled events. I had preregistered long in advance so I assumed there had been some clerical error. When I looked around and noticed that none of my friends and colleagues from major environmental organizations were there, I checked the list again. Their names were not listed either. It was obvious that something fishy was going on. I had no idea what that might be, but I knew I hadn't come all the way from California to miss the whole thing.

Sitting on a park bench facing the sidewalk that circles the castle grounds, I pondered my dilemma. I noted that on the ground level there was a huge ballroom where the delegates were enjoying the opening-night cocktail party and musical performance. The castle's large French doors were open to the outside. Between me and the party was a hedge about five feet high. Police with dogs were guarding the castle, patrolling in a circular fashion. I assessed how far apart the police officers were from one another and how fast they were walking. Then I made my move. Using my training as a high school hurdler, I got a running start, high-jumped, and rolled over the hedge. Landing on my hands and feet, I brushed off my dark blue suit and tie and ran like hell into the ballroom. I could hear police whistles and barking dogs behind me, but I easily blended in with all the blue-suited delegates.

I quickly found Russell Train, the head of the US delegation, a leading environmentalist and a good man. We had worked together on several projects, and I knew him pretty well. In the days when many

Republicans were environmentalists and their definition of "conservative" included conserving our natural resources, he had founded several wildlife protection organizations. At the time of the conference, he was the administrator of the US Environmental Protection Agency. When I told Russell what had happened, he quickly got me credentials, making me the *only* environmental nongovernmental organization representative in attendance. I set out to make the most of my unique opportunity and started talking to as many of my fellow delegates as humanly possible At some point, I recognized Laurance Rockefeller, who had been so generous and helpful in the acquisition of the Seven Sacred Pools in Maui. He was having a cup of tea in the lobby, and I was happy to join him for a short conversation.

Later that evening, I met Margaret Mead, the world-famous anthropologist and author. She knew exactly what was going on: seven powerful European countries, along with the United States, had formed a secret alliance to take the teeth out the UN conference agenda so it wouldn't pose a threat to their economies or major corporations. Calling themselves the Brussels Group, representatives from Belgium, Britain, France, Germany, Italy, the Netherlands, and the United States – as well as the powerful International Chamber of Commerce – met surreptitiously several times before the Stockholm conference began, eviscerating the ambitious goals environmentalists had hoped to accomplish by excluding NGO leaders like myself. We know all of this now because of the British "thirty-year rule," which, much like our Freedom of Information Act, requires that confidential government documents be released to the public after thirty years.

In the meantime, the other environmental activists who were barred from the conference came up with a brilliant and subversive idea. Led by Stephanie Mills – who would become a well-known environmentalist, prolific poet and author, and assistant editor of the *Whole*

Earth Catalog's Co-Evolution Quarterly – they rented a lovely house near the castle and held "salons" every evening, using the term coined by seventeenth-century French intellectuals to describe an intelligent and sophisticated exchange of ideas. With Margaret Mead as their big-name draw and Stephanie as the queen bee hostess, they invited delegates from the conference to their nightly salons, ensuring that the true environmentalists' point of view would be represented in the next day's conference discussions.

The salon idea was so successful that Stephanie continued to host them in San Francisco when she returned home. She rented a local house and decorated it beautifully, with ornate gold leaf on the ceiling. (I guess Stephanie's tastes have changed because she now lives in a log cabin in northern Michigan.) She invited executives, politicians, religious leaders, and other leading figures and made it clear to them that anything they said would be held in confidence. That allowed for lots of frank and freewheeling conversations and gave us a chance to get across our ideas to people with influence.

I'm still struck by Stephanie's improvised creativity in Stockholm. Banned from the conference she was invited to attend, stranded in a foreign city with no local connections, she somehow crafted a response so interesting and delightful that everyone rushed to her venue to learn and talk and network. She figured out how to bring the mountain to Mohammed, a very useful talent in the world of environmental activism.

The conference accomplished its goal, and a second conference was scheduled to vote on whether the UN Environmental Programme (UNEP) agency should be officially established, and if so, where it would be headquartered. Assuming the UNEP would, like so many other UN agencies, set up shop in Geneva, the delegates from the Western nations that dominated the UN were surprised to see a sea of black hands pop up in the audience. This was a time when former colonies of European countries were asserting their place on the world stage. They demanded the agency be headquartered in Nairobi, Kenya. Although the Western delegates were shocked at first, they cynically warmed to the idea over time, figuring that Kenya's distant location and limited infrastructure would dilute UNEP's power. They were correct, unfortunately, but I'm still optimistic that the agency will grow in influence in the twenty-first century.

One of those raised hands belonged to Kenyan environmentalist and future Nobel Peace Prize–laureate Wangari Maathai. Brilliant and charismatic, sophisticated and down-to-earth, Wangari would play a major role in world environmentalism and an important part in my life. Like another hero of mine, Aldo Leopold, Wangari was a person of firsts – the first East African woman to be awarded a PhD, the first woman to chair the Department of Veterinary Anatomy at the University of Nairobi, and the first African woman to receive the Nobel Prize. Those are only a few of her accomplishments. Throughout her life, she fought for human, political, and environmental justice, believing that all three were inextricably connected. Best known as the founder of the Green Belt Movement, which helps local women plant trees as an act of environmental and political empowerment, she gained international prominence in 1989 fighting against the development of a proposed sixty-story skyscraper in Uhuru Park, one of the few open spaces left in Nairobi. Today, Green Belt has planted more than forty million trees in Africa, and I'm proud that my nonprofit, Resource Renewal Institute, was able to help Wangari expand Green Belt into many other countries.

Although her life was threatened many times by the Kenyan dictatorship, she died of natural causes in 2011, far too soon for the world she worked to protect and for those of us who loved her. At her San Francisco memorial, one of several held worldwide, Vice President Al Gore described "her courage as almost unimaginable."

There's a perfect punchline to my experience at the 1972 UN conference in Stockholm, one that was nearly thirty years in the making. In 2001, UNEP, the agency that refused to seat me and forced me to high-jump past police guards to get inside, awarded me its highest honor, the Sasakawa Environment Prize. I was utterly surprised that it included a $200,000 award and was deeply honored to receive it.

Thinking back to the charming and creative way Stephanie Mills responded to our exclusion from the Stockholm conference reminds me of how important communication is to winning people over to the environmental cause. Coincidentally, a few months before Stockholm, I discovered perhaps the greatest piece of environmental advocacy ever written – *The Lorax*, a 1971 children's book by Dr. Seuss. A simple fable of greed and resource management, *The Lorax* has gone on to sell more than two hundred million copies and was turned into a film, a television special, and a stage musical. I was so affected by the book when I read it that I looked up the phone number of Theodore Geisel, the real name of its famous author, and gave him a call to say thank you. He was a pleasure to chat with, and after a few minutes, he invited me to visit him at his home in La Jolla, a coastal village north of San Diego. I couldn't believe his generosity and my good fortune. I drove down the coast and we swapped stories and drank cocktails.

Since nothing I've ever written captures the truth and urgency of environmental preservation as well as *The Lorax*, I've included the book in most of the courses I've taught over the years, including those at Harvard and Berkeley. I even solicited the help of a student whose grandfather had refused to talk to me about a rare wetland area he owned, one I was trying desperately to save. Home for the holidays, the grandson gave his grandfather a copy of *The Lorax* for Christmas. The land soon became a sanctuary. But what stands out most is the memory of reading the book to my two grandsons when they were young. I read it to them so often they knew every word of it by heart.

Often in large organizations, the culture changes slowly and imperceptibly, until one day you realize you don't work at the same place anymore. That was not the case at The Nature Conservancy. Our organization was totally and unexpectedly transformed in the course of a few hours, at a board meeting.

It all started when simmering differences between two factions on the board – the Pittsburgh liberals and the conservative New York corporatists – exploded into an all-out fight. One of the liberals proudly announced that he had landed a $500,000 grant from the Ford Foundation. A right-winger named George Cooley went ballistic. He thought the Ford Foundation was run by a bunch of communists and demanded TNC refuse their money. Everyone started yelling, and within a few minutes, Cooley and five of his supporters had resigned and walked out of the meeting. They were followed by my wonderful boss, Walter Boardman, who resigned in disgust. Walter was a soft-spoken, gentle person who had no patience for boardroom fireworks. As he was leaving, he told those left in the room that they should name me executive director.

Looking back on the whole thing, I find it quite sad. George Cooley may have held extreme political views, but he was a good man. (He had a dry sense of humor and once told me that I was the only environmentalist he would shake hands with.) The truth was, George had been blindsided by the Ford Foundation announcement, something he had a right to know about long before the meeting. Beyond that, the board members should have been able to hold different viewpoints and still work together.

When the smoke cleared, the board and staff followed Walter's suggestion and offered me the executive director position. They added the presidency to sweeten the deal.

I turned them down. As far as I was concerned, I already had the perfect job. In five years, I had preserved hundreds of natural treasures across the West, building my thirteen-state region from a nonentity into a major land-saving force. My young family and I loved our life in the Bay Area where I could spend my work and leisure time in nature – hiking, fishing, and duck hunting with clients, colleagues, and friends. Best of all, I was three thousand miles from the bureaucracy and politics of TNC's D.C. headquarters, running things as I saw fit.

The board made a counteroffer. Would I serve as acting president for three or four weeks until they found someone permanent? I agreed, and my wife and I and two little ones moved into an apartment in Washington. Somehow a few weeks turned into almost a year; I began to wonder if they were serious about looking for a permanent executive director.

During those months that I was living and working in Washington, I was lucky enough to forge an unlikely and meaningful friendship with Supreme Court justice William O. Douglas. Nicknamed "Wild Blue" Douglas for his environmentalism and love of the outdoors, he called me out of the blue to meet with him at his office in the Supreme Court building. He explained frankly that he was about to ask his wife for a divorce and wanted to donate a beautiful property he owned before it could become part of their divorce settlement. Once the land was saved, we continued to meet at his office and talk. His philosophy of life had a great effect on me. He had clear and strong core beliefs that made him, as he described it, "impervious to criticism."

My friendship with Justice Douglas coincided with the famous "one man, one vote" Supreme Court decision mandating that electoral districts be apportioned by population, with each district having roughly the same number of people. The ruling meant that urban areas would instantly acquire more political power. It made me think about a recent experience I had had in San Francisco's Washington Square, a lovely park a few blocks from Chinatown. I came across a group of older people performing their morning tai chi, the slow and graceful Chinese martial art. The friend I was with said to me, "This park is their living room."

This park is their living room. This park is their living room. The phrase lodged itself in my brain, and an idea began to emerge. With the impact of "one man, one vote," what if TNC were to set up a division dedicated to creating and saving urban open space, making cities more beautiful and livable?

I presented the idea to TNC. They loved it. I set out to create a new division and a new position for myself. But the more I thought about it, the more concerns I had. TNC had started a number of worthwhile new divisions – protecting hawks and hawk habitat, saving marshes, and so forth – and, by necessity, they would all be competing for priority and funding. I started to wonder if my exciting new idea would ever see the light of day.

Around the same time, I received a letter from my friend Charles Borden, the old sailor and writer who had donated his ranch on Spindrift Point.

> ....*Days are fair and warm here now. More "footsteps-of-spring" are out along the paths and Ceanothus "sierra blue" are bursting their buds. Looking out – as I type this, directly below here a quarter-mile long finger of spindrift is edging south in the tide flow toward the Golden Gate. On these first spring days everything here seems ageless and deep still – rocks, earth, sage, chaparral broom, cormorants and beneath it all so wonderfully alive. The heavy rains of January have brought forth stirrings of new life everywhere. The air, since the last rain ended three days ago, has been fresh with new smell of lichen, wild lupines, grasses and earth mingled with the salt of sea air in the early morning onshore breezes. Nights are sharp and deep still. Yesterday the bay from here to Pebble Beach*

was filled with over 1500 western grebes (Eleanor counted them for her daily log). Early last evening about 200 migrating red throated loons circled in for the night and are still here. Also a small group of surf scoters and a few greater scaups.

Hope you are making definite plans to return. Duties are just that – men have been coming and going and responding to them for centuries and are little the better for it. Wisdom lies in not rationalizing or waiting too long. Hurry back to the things that really matter.

Charles Borden

Charles's letter struck me right in the heart. I realized that I had compromised too much of myself in the last year. I belonged in the beautiful Bay Area near the ocean and wildlife and open land. I needed to break free from a way of life I knew wasn't good for me anymore. I set out to be like William O. Douglas – impervious to criticism. I told the people at TNC that I was leaving Washington and returning to my old job in San Francisco. It was clear to them I meant business, and within a few weeks, I was back in California.

There is a great exhilaration, almost intoxication, to the business of saving land. You are rewarded with a tangible and beautiful piece of earth that nourishes animal and plant life, that anyone and everyone is entitled to experience, enjoy, and study.

But there is a less exciting part of conservation work that may be even more important – the defense of places already saved.

The inexorable pressure to sell, to develop, to change established trusts and laws is a never-ending counterforce in environmentalism. More and more, we are relaxing areas legally committed to preservation "in perpetuity." Such reversals involve a loss of integrity and honor, a betrayal of the pledges made to generous, kind-hearted people, often at the end of their lives. Frankly, I feel tremendous guilt whenever my promise to them is breeched by forces out of my control.

The Nature Conservancy is a great organization with huge accomplishments. I was there when it grew to national prominence and admire much about it. But I think my years with TNC and my lifelong experience as an environmentalist give me the right to offer constructive criticism as well. To me, TNC has aligned itself with corporate America at its own peril. Its current executive director is an executive at Goldman Sachs and once served on the board of the American Petroleum Institute. So it's not surprising that TNC leases some of its land for economic reward. It earns income from oil wells on land it has been charged with preserving. Clearly, TNC has lost touch with its original purpose.

True, a business background can be a valuable tool in managing institutions, yet history shows us that it does not necessarily go hand in hand with integrity. My sincere hope is that TNC will rescue itself from what amounts to a corporate takeover by recalling its storied past and restoring balanced management.

The British National Trust offers TNC and all of us a fine model to strive for. In the sixties, a crashed economy put enormous pressure on BNT, requiring big changes. It had acquired and kept so much land that more and more of its annual budget was needed for management expenditures. The staff was intent on continuing with the fun part of the work – acquisition – and passing off acquired lands for others to manage. The board chose otherwise, to cut back on acquisition and to carefully manage the places they already owned. Because it built on an existing heritage, BNT got the support of local communities throughout the UK. Today, the idea of selling off BNT properties in Britain is unthinkable, even to developers. In addition, the trust's proven reputation as a responsible land manager resulted in Parliament providing it with tax advantages and a yearly government stipend to use as it wishes. Today, BNT manages the great historic estates of England and is a trusted consultant on British environmental policy.

In much of England and Wales and most northern European countries, the heritage of open space is reinforced by an age-old custom, sometimes written into law, called the "right to roam," or "every man's right." In these countries, private property is subject to public access for hiking and short-term camping. As a result, people grow up with a more nuanced distinction between public and private land, and a natural respect for the value of free and open space.

My first encounter with the "right to roam" was in Sweden, and it hit me like a ton of bricks. I realized that while I had spent most of my career in the pursuit, negotiation, purchase, and politics of acquiring private land, turning it into public land, then protecting it from private takeover, the Scandinavians enjoyed access to public and private places as a birthright. No lawsuits. No court fights. No sales pitches to wealthy prospects.

Lesson learned: There's always a better way to do things.

Probably the best example of a successful defense of open space in the United States is the Grand Traverse Regional Land Conservancy in Michigan, which protects forty thousand acres and 124 miles of river, lake, and stream shoreline. As with the BNT, the GTRLC earned its success and community respect by bringing together government, industry, and NGOs to forge a joint agreement. The nonprofit land trust, collaborating with the state and hundreds of volunteers, has advanced hundreds of projects in twenty-five years, half acquisitions and half easements. At this point, I doubt any developer would try to violate its lands. With both the BNT and the GTRLC, enough time elapsed and sufficient community education took place to grow a necessary base of support.

For a long time, I was critical of land trust managers; I thought they weren't aggressive enough in fighting off threats to sell off or commercialize the places they were entrusted with protecting. After a while, I realized that land savers and land defenders are usually very different kinds of people. Some, like me, are natural savers: people who enjoy getting to know owners, who are happy to become a part of their lives, who want to work with them to realize their wishes for the land they love. The defenders, on the other hand, have a little pit bull in them. They like to go toe-to-toe with their opponents to make sure public land stays public. Over time, as public lands have become increasingly imperiled, I've learned to become a defender as well as a saver.

My point is, when land defense is done well, when it's supported by stakeholders like the government, nonprofits, and the community, when it's held up as an intrinsic part of local and national heritage and values, it can be hugely successful and last forever.

So what's my advice to TNC? Return to your roots.

Remember your pledge to thousands of generous donors to preserve and protect the land they cherished and intended for the enjoyment of generations to come.

Don't take money for easements on land that is already saved and you intend to keep.

Don't seek or accept government funds that invariably come with strings attached.

Don't mismanage properties you are responsible for restoring, and misdirect income back to your headquarters – as you evidently have done in California's Staten Island wetlands.

And especially, don't get into bed with oil companies. You have chosen to let them drill on protected land, which earns TNC millions in dirty money.

Instead, remember how you became a great organization in the first place. Cherish your generous land donors, raise funds from caring individuals, enlist the participation of volunteers, engage with the communities you service.

And always think of the great people who put their faith in you.

People like Elting Arnold, corporate counsel for an international bank who volunteered long hours helping TNC with its legal needs. Or the unforgettable physician, Albert Byrne, who lived in a rundown shack in Texas. He had been tortured as a volunteer for the Abraham Lincoln Brigade in the Spanish Civil War, then became a doctor on an Indian reservation. At the end of his life of giving to others, he entrusted me and TNC with a fifty-acre ranch in the Los Altos Hills, on the San Francisco Peninsula. I promised him we would use our last breath and last dollar to defend his gift forever. Years later, I did. As that part of the peninsula grew into Silicon Valley, the land increased in value from a thousand dollars an acre to a million. The city fathers wanted to sell the much-loved park to build a world-class city hall and bury the area's phone lines. My strategy was to enlist the town's children, who used the park every day, to form an irresistible lobby to save the park. The children put out a flyer challenging the council's idea to develop the park and delivered a copy to every home. They were soon joined by their parents and Los Altos' older citizens who had volunteered to work on the park when it was first founded. The park stayed, and my promise to Albert was kept.

Never give up. Never give in.

CHAPTER SEVEN

Lesson Learned | Drink Forty Cups of Tea

I settled into my old job in San Francisco, but it wasn't the same. TNC went through a series of short-lived executive directors who micromanaged my every move. It was clear I would never again have the independence that had made working at TNC such a pleasure and allowed me to do my best work.

The time had come for me to go my own way, just as I had when I left the Visking Corporation and the University of Michigan PhD program. True, I now had a young family to support, but I knew that if all else failed, I could always make a good living as a salesman. Besides, I had pretty much been my own boss for all those years at TNC, so I felt confident about taking off the training wheels and starting my own venture.

I had a clear vision of what I wanted to do next. My friend had said it all that foggy morning as we watched elderly Chinese people perform tai chi in San Francisco's Washington Square.

"This park is their living room."

With those six words as my north star, I set up shop as The Trust for Public Land at Eighty-Two Second Street in San Francisco's Financial District. I dedicated myself to creating and saving urban open space and making cities more beautiful and livable. I wanted my organization to be small and agile, without bureaucracy or hierarchy. And I had a few guideposts I wanted to follow. I still think they're good rules of thumb to help any small environmental nonprofit succeed with integrity:

Get local communities involved.
Peer relations save land. Whether it's a park, a community garden, a playground, or a trail, people have to feel they have a stake in a plot of land in order for it to thrive. The arrangement can include actual ownership by the neighborhood as a 501c3, but it doesn't have to. What's most important is that everyone participate in some way – planning, development, building, maintenance, activities. The place should be a part of their lives.

I first saw this approach flourish at my beloved Carnegie Library in the small Michigan town where I grew up. Virtually all of the 2,811 libraries Carnegie endowed across the United States and other countries followed "the Carnegie formula." Any town requesting a library had to provide its own building site, put up 10 percent of the money needed to build and maintain the library, and prove it had the means to pay a staff once construction was completed. The formula also required that public as well as private funds be used to run the library, and that it always be free and open to all. In other words, everyone contributes and everyone benefits. As one of its beneficiaries, I can only say that without my Carnegie Library, I would never have gained the knowledge or understanding to pursue the life I have.

Buy the land, then give it away.
I believe strongly that acquiring *and* managing public land is a fool's errand. You can't do both well. In my opinion, trying to do it all has been a major source of TNC's problems. The British National Trust, on the other hand, made the difficult decision to stop acquiring land and concentrate on managing its vast holdings of historic estates and open space, which it does beautifully. For my fledgling nonprofit, I chose to do the

opposite – acquire the land, work with the community to set up a solid ownership structure, then move on. In and out. Of the thousands of acres my team and I saved in my years at TPL, I estimate 80 percent of the projects turned out well over time. I think that's a pretty great average, and I would handle it the same way if I were starting out today.

Include an element of self-interest for your donors.
The tax-saving formula Putnam Livermore helped me work out at The Nature Conservancy had proven itself to be extremely attractive to contributors, so I decided to make it TPL's primary fund-raising tool. I determined that my target audience for donations would be wealthy people who loved open space and believed everyone had a right to enjoy it. I thought about where these individuals tend to congregate – duck clubs, hunt clubs, fly-fishing resorts – and began frequenting these places myself. It worked. Not only did I find the money I needed for TPL, I developed friendships I've enjoyed ever since.

Get dollar-a-year people involved.
You don't hear this expression much these days, but the concept is as valid as ever. "Dollar-a-year" refers to outstanding executives or professionals, often retired, who are financially secure enough not to worry about money anymore. I relied on their involvement to get TPL off the ground and keep us going. To a person, they were grateful to contribute their talents and expertise to our cause. When it came to lawyers, I always sought out the ones who commanded or had commanded the highest hourly fees. I knew they were the best at their trade, and fortunately, they rarely charged us at all.

I remember one dollar-a-year man in particular. Richard Goodspeed. A charming San Franciscan with a British sense of style, he had been president of Lucky Supermarkets in Northern California. Richard walked into my office one day, asking if there was anything he could do. From that day on, he was a fixture at TPL. His intelligence and business acumen took our operation to the next level and helped us acquire many properties in Oakland and beyond. He also taught me the value of working with potential donors for as long as it takes to win them over. His endless patience was invariably rewarded. He used to say that if you have to drink forty cups of tea with someone to seal the deal, then drink the forty cups of tea. It's a symbol of perseverance I think of often, and it usually works.

A while later, Crayton Peet, past president of Safeway, joined his former rival from Lucky to volunteer for us. His contributions were as significant as Richard's. For some reason, both men loved cut flowers so I made sure there were always bouquets on their desks when they arrived at work. It was the least I could do.

Another excellent dollar-a-year man was Sy Foote. As a young IBM employee, Sy had an idea that made him financially independent for life. While everyone else was focused on the company's data-processing technology, Sy noticed that the ever-present cardboard punch cards were piling up all over the office and making a mess of everyone's desks, tables, and cabinets. So Sy left IBM to design a line of cabinets specially made to hold and organize punch cards. First he sold his products to IBM, then sold the entire business to a large office furniture company. With no need to earn a living, Sy, like Richard Goodspeed and Crayton Peet, decided to devote himself and his remarkable business skills to TPL. As if that weren't enough, Sy also left a generous bequest to our organization when he died.

Then there was Lewis Reid, who brought a breadth of experience to TPL. Lew is a lot of fun and always has a twinkle in his eye. We liked to play practical jokes on each other, although his were generally more successful than mine. Lew started out as a corporate lawyer who soon discovered that he hated corporate law. He shifted gears, doing legal work for large foundations and nonprofits like Blue Cross, then found his way to Washington, D.C., where he was legislative aide to California senator Thomas Kuchel, a moderate Republican. While in Washington, he slipped through a monumentally important addition to a routine environmental policy bill, essentially inventing the mandate requiring an environmental impact study for every public and private project. After leaving Washington, he worked in state government in Sacramento.

These experiences, and his own intelligence and creativity, made him a savvy political operative. When I met him through our mutual friend Putnam Livermore, he was just coming off a huge victory. A state proposition he had worked on was approved by the voters, creating the California Coastal Commission, an institution that has saved most of our thousand-mile coastline from development and exploitation. To get the proposition over the top, he employed the most inventive tactics imaginable, including this one: he bought his young son one share of Pacific Gas & Electric stock. Then the boy stood up at a rally and announced that he was an owner of PG&E and was upset that his company was lobbying against something as important as a coastal commission. That night, it was all over the news.

Don't have members.

Many nonprofits work on the membership model. By paying an annual fee, members gain certain rights of participation in the organization. It's a tempting way to raise money, but things get complicated very quickly. Don't get me wrong. I love volunteers and could never have succeeded at any nonprofit without their support and commitment. But members are different. They expect to have a voice in policy decision-making, and rightly so. It's a simple case of too many cooks. So I did not solicit members when I started TPL, and things ran smoothly as a result.

Hire extraordinary people and help them grow.
If your goal is to spread environmentalism far and wide, hiring the right people is the way to get there. At TPL, I chose my staff with extreme care – and a good dose of intuition. I didn't demand a perfect résumé or even direct experience. I was looking for three attributes: a curiosity to learn, a personality that projects confidence, and a commitment to a life of purpose. If they had those qualities, I was happy to teach them what they needed to know. And as is often the case with being a mentor, I learned a great deal from them as well.

Evidently, the hiring business is something I'm good at because the group that worked for me during my five years at TPL has gone on to found or lead a slew of worthwhile environmental organizations themselves.

I started out with a crackerjack staff of twelve. One of my first employees was a wonderful character named Tom Macy. He had been an air force pilot in Vietnam, then switched to the army to become an infantry platoon leader, which he still considers the best time of his life. Tom's idea of fun was to join his brother on cross-country skiing and camping trips along the frozen Yukon River in Alaska. They would eat whole sticks of butter to stay warm on the trail. Obviously, Tom was a person who wasn't afraid to take risks. He jumped into TPL with boundless enthusiasm. Like me, he knew the power of salesmanship and used it to acquire a bundle of places on our behalf. He was also great at being what the business world calls a sales manager. He taught our younger, inexperienced staff how to work effectively with prospective donors and then close the sale.

In 1985, Tom cofounded the Conservation Fund, which has saved over a quarter of a million acres of working ranches, river corridors, and outdoor recreation areas in Colorado. He's still a part of the organization, currently serving as its western representative for conservation acquisition. His cofounder, Pat Noonan, is also an outstanding pioneer in the land-saving movement and is chairman emeritus of the Conservation Fund.

Another one of our stars at TPL was Steve Costa. Steve had an impeccable pedigree as a community organizer. As a young man, he was a volunteer with Vista, the "domestic Peace Corps" that was part of Lyndon Johnson's War on Poverty. Later, he was trained by protégés of the legendary community advocate, Saul Alinsky, and wound up working in Oakland for a group of churches that pressed for neighborhood rights and betterment. That's when I met him. Steve had the local understanding, close relationships, and on-the-ground knowledge I lacked and contributed greatly to our efforts.

Soon, two young lawyers, Steve Steinhour and Greg Archbald, joined our staff. Unlike the older, financially secure dollar-a-year men who volunteered at TPL, Steve and Greg were employees, working for far less than they could have earned at a law firm or corporation. I first got to know Steve at The Nature Conservancy, and later when I left TPL to work in Jerry Brown's cabinet, I hired him to implement a bond measure whose monies were to be used to acquire private lands for public parks and open spaces.

A Midwesterner who graduated from Stanford and Yale, Steve is a terrific lawyer who, like Lew Reid, hates practicing law. Lucky for me. At TPL, we were inventing a new kind of nonprofit that required us to negotiate with major banks, companies, and local governments. It was uncharted legal and financial territory, which Steve navigated skillfully. He wasn't afraid to take on Bay Area power players either, and together, we did our share of stirring things up. We both focused single-mindedly on results. Steve never discouraged me, as many attorneys do, from taking calculated risks, and he always worked like hell to make sure everything we did was airtight and buttoned up.

Our talents seemed to dovetail perfectly. He describes himself as someone who is not a leader, but the best Number Two person you'll ever find. So I took risks, and he assessed and managed them. I used to call Steve my consigliere. Which is exactly what he was. Again, lucky for me.

Greg was just as effective. He had been a lawyer at a prominent firm when a client asked him to help her save a piece of land. In the course of doing research for his client, he found me – and his life's calling. He left his lucrative practice, joined our staff, and spent the rest of his life working in the movement.

Greg uses a lovely phrase to characterize his environmental philosophy – "caring from the heart." It's how he engaged thousands of active volunteers in preserving the Golden Gate National Recreation Area. As Greg explains it, institutions like churches and civic organizations last from generation to generation because they forge and maintain traditions. The same is true for public lands. Every time a class of school children, a family, or a senior citizen gets involved in cleaning up trash, pulling weeds, or planting seeds, they become what Greg refers to as "built-in ballast." Which means they are committed to protecting a place in perpetuity, the goal of every land saver.

Greg has the numbers to support his theory. By the time he retired from the GGNRA in 1999, there were thousands of volunteers helping out. Decades later, he still has a passion for nature. Today he raises roses for the public gardens of Nevada City, California. When his fellow citizens have something to celebrate, Greg's roses decorate the tables.

Unlike veteran pros like Tom, Steve, Steve, and Greg, John Nelson was a young guy with almost no experience at all. I met him through the San Francisco Zen Center where he was studying to become a Buddhist priest. I knew that the center's curriculum included a public service component, and I hoped my nonprofit on a budget could get some volunteer help.

The idea occurred to me because I had once engaged a young man to be a caretaker for some land we were holding at TNC. He and his girlfriend lived rent-free in exchange for overseeing the property and keeping it in good shape. I didn't know then, but my little idea turned out to be the beginning of a national movement. Today there are people all over the country who live free on public lands by working as caretakers.

Unlike the young couple at TNC, John Nelson became much more than a caretaker. From our first meeting, I could tell he had an active mind and a commitment to social justice, having organized a labor union in his early twenties. I'm not sure why or how, but soon after his arrival at TPL, he became fascinated with land trusts, a concept that had been around since the 1890s but not used much in the West.

The first American land trusts were created as a way to preserve palatial East Coast estates built by wealthy industrialists in the late nineteenth and early twentieth centuries. In a time before antitrust laws and income tax, these men amassed fortunes they could never spend in their lifetimes. Of course, that didn't keep them from trying. As years went by, many of these homes became too expensive to maintain and fell into disrepair. That's when citizens of places like Newport, Rhode Island, and Long Island and Hudson Valley, New York, devised the land trust – which is simply one party holding a piece of land for another party's benefit – to restore and repair the estates for historic preservation and public enjoyment. The trusts also provided the heirs of Rockefeller, Vanderbilt, Carnegie, and Duke attractive tax deductions for donating their property.

About the same time land trusts took hold in the eastern United States, the British National Trust embraced the concept and made it their own. As I describe in Chapter Six, it became one of Britain's greatest institutions, a source of pride for its citizens and a powerful connection to the past.

As John Nelson and I talked and brainstormed, we both agreed that introducing land trusts to the West was a terrific idea. It integrated perfectly with Putnam Livermore's tax benefit strategy for land donors developed for The Nature Conservancy, and it appeared to be the perfect tool for saving urban spaces as well. We began by establishing the Sonoma Land Trust and Napa Land Trust. Both have operated successfully for more than forty years. When you consider the pressures put on Napa and Sonoma from today's wine, tourism, and real estate interests, those trusts have proved more important than we even imagined.

When I founded Resource Renewal Institute a decade later, John Nelson worked with us on our seminal idea of Greenplanning, acting as liaison between RRI and the Dutch government. John continued to use his financial savvy for good. As managing partner for Wall Street Without Walls, he teamed financial industry experts with community development groups and connected them to capital marketing and financial institutions.

There were many other exceptional folks who contributed their talents to TPL. Like Pete Stein who showed up at our office at age twenty-two and went on to run TPL for the eastern United States. Or Jenny Gerard who performed outreach for us, speaking to hundreds of community organizations on the arcane but vitally important subject of property easements. Or Phil Wallin who originally came to me for a community-oriented job that would qualify him to avoid the draft. What he lacked in experience he made up for in confidence and intellect. Phil, who soon became a committed land saver, took a break from TPL to attend law school at University of Chicago so he could return to environmentalism armed with greater skills and knowledge. Before he left for school, Phil casually mentioned to me that while he was in

Chicago, he would take some time out from his studies to write a manual for land savers. I said, "Great, thank you," and never thought about it again. Several months later, Phil showed up at our office and dropped two enormous tomes on my desk. It was a complete guide to every aspect of our new profession. After fifty years, it's still a relevant and useful tool. Later, Phil moved to Portland, Oregon, where he founded a river-saving nonprofit, Western Rivers Conservancy.

Train your staff – but also your donors and recipients.
John and Phil were two of several young people I hired, all of them bursting with potential. They wore the uniform of the day – long hair, sandals, jeans, and army surplus jackets. In those outfits, they were never going to convince bankers and land owners to work with us. So I marched down to Goodwill and bought them a bunch of used Brooks Brothers suits and dresses for $5 apiece.

Next I held intensive training sessions on everything from resources and carrying capacity to urban land trusts and participatory democracy (what I call "50 percent plus one"). Before anyone received a paycheck, they were required to do a five-minute presentation. Presentably dressed and armed with a sales pitch, they were ready to go out into the world.

Maybe even more important was our custom of training not just our employees but also our land donors and recipient organizations. For donors, our training highlighted both the tax benefits their land donations would offer and the priceless human value of their generosity – playgrounds for children, community gardens growing fresh organic produce, open space in crowded urban neighborhoods. For the recipients, training provided essential tools for managing the precious asset they were entrusted with. Its focus was on leadership, neighborhood decision-making, and training employees and volunteers to protect and maintain the land.

The biggest believer in my approach to training and management was and still is Steve Steinhour. To be honest, I'm not aware of any special techniques I've used to bring along those who've worked for me over the years. As I mentioned earlier, I've found that if you hire the right people, all they need is respect, the opportunity to learn, and a good challenge. In general, I try to lead by example. Of course, it doesn't hurt to unobtrusively monitor their performance once in a while to make sure they're on the right track. They almost always are.

Steve likes to talk about my habit of including new employees in important meetings and introducing them to clients by recounting their many fine qualities. These introductions served to earn the client's trust, but even more important, it bolstered the confidence of the young TPL staff members. After all, if I believed in them, they had every reason to believe in themselves.

Steve says he took my approach to training and management with him to Sacramento. He describes it as flexible, tough, firm, and creative. He says I can be polite, brusque, or outrageous depending on which behavior is most effective in a given situation. For example, Steve remembers meetings with adversaries who assumed I was there to argue and push back. Instead, I came equipped with an entirely unexpected new solution. It usually disarmed them and led to success. Steve used a similarly creative approach in Sacramento. Rather than fighting vainly against the intractable and entrenched government bureaucracy he had inherited, Steve simply transferred the whole crew out of his department and brought in others to do what was supposed to be done.

I think Steve gives me far too much credit, but I very much appreciate the compliment. I feel the same way when he refers to me as "his second father," a lovely thing to hear.

Have lunch together.

At TPL, I started a practice that reaped rewards there, and at every other nonprofit I founded since then. It's as simple an idea as it is effective: treat your staff to a deli lunch at the office every day, and, whenever possible, invite a variety of knowledgeable guests to join in. First of all, it's good for morale and gives everyone a chance to be friends. Next, it lets them talk about what they're working on and percolate new ideas, which makes everyone better at their job.

When guests are there, the atmosphere is more relaxed and natural than it would be in a conventional office meeting, leading to more good ideas. And if the guests happen to be former employees, they act as organization historians, sharing stories that give the current staff a sense of continuity.

In the earliest days of TPL, I couldn't even begin to assemble a team until I raised some money. I set out to win a large grant that would get us off the ground. I prepared a proposal for the Ford Foundation describing what I wanted to do. With persistence – otherwise known as forty cups of tea – I somehow scored a meeting with the foundation at their offices in Manhattan. I met with a group of employees, each one a specialist of one form or another.

I presented my idea. Before anyone else had a chance to comment, the foundation tax accountant glared at me and said, "This is foolish and stupid and will never work. A waste of my time." Then he got up and left. The others in the room were deeply embarrassed and tried to say nice things to make me feel better. But it was clear that I had lost.

I went out the door into a cold, dark winter day, as despondent as I have ever been. It was a long way back to California. Back home, I reviewed my ideas and ran them by some fiscal experts. It still made sense. I decided to go right back to the Ford Foundation.

Somehow I got them to agree to see me again. On my flight to New York, I remained fearful.

When I arrived at the foundation, I learned that the financial person who sabotaged my first presentation was not liked or respected at all. Plus, he had a feud going with the environmental branch of the foundation. When I entered the second meeting, the same guy was there again. This time I confronted him and made my case.

I got the $250,000 grant I needed. Not only was that grant the basis for TPL's success, it was yet another example of the power of persistence. As bad as I felt after my first meeting, I knew my plan was a good one, and I forced myself to approach the foundation again. I had nothing to lose and wound up gaining everything.

The next challenge was to find TPL's first project. We were hoping to acquire something spectacular and high profile to establish our reputation and set a path for everything that followed. The gods must have been smiling on us because Greg Archbald received a phone call from Richard Thompson, a real estate agent in Los Angeles who was familiar with TNC. He explained that a prominent attorney, John O'Melveny, had called him with a proposition. He and his wife owned a beautiful ranch, nearly seven hundred acres in size, in the San Fernando Valley, and she had passed away. Both valued open space in an area where it was disappearing quickly, and Mr. O'Melveny wanted the ranch – also known as Bee Canyon – to be preserved. If a conservation organization would buy his wife's half for a very reasonable price, he would donate his half in five yearly installments. Another stroke of good luck followed. As an attorney, Mr. O'Melveny knew and respected Greg Archbald's father, also a Los Angeles attorney, and his grandfather, who had been a superior court judge. The fact that Greg was their son and grandson convinced Mr. O'Melveny that the new and untested Trust for Public Land could handle the project.

We worked out a way to turn the ranch into the second-largest park owned by the city of Los Angeles. It was a highly complex transaction for a tiny organization just getting started, one involving the Los Angeles Department of Recreation and Parks and the city council. But Greg did an excellent job and pulled it off. What is now O'Melveny Park became our first acquisition.

It was just what we needed: a magnificent open space preserve near an urban population that would become a public park for everyone's enjoyment. It was also a big, newsworthy story that legitimized TPL and put us on the map.

Around the same time, I did something risky. It involved the old Wilkins Ranch, a wonderful piece of oceanfront property forty miles north of San Francisco. Its nearly fourteen hundred acres of coastal hills overlooked Bolinas Lagoon and the Pacific Ocean. Nicholas Charney, the founder of *Psychology Today* magazine, had bought the old ranch with the intention of turning it into kind of a commune. Charney lived elsewhere but allowed a number of young people to occupy the land, raise organic vegetables, and live in the barn. Eventually everyone abandoned the place. Charney decided to sell and called me with an offer.

Like Bee Canyon, Wilkins Ranch was a lovely place for urbanites to enjoy nature, and a perfect example of what TPL was all about. However, even with our quarter-million-dollar Ford grant, it was too much for us to spend. Still, I was convinced that our first acquisition in the Bay Area would determine whether we would survive as an organization. Somehow we made the numbers work, and before long, the National Park Service bought the land from us. It soon became an anchor parkland of the new Golden Gate National Recreation Area.

With our one-two punch in LA and the Bay Area, TPL was off and running.

Next, we came up with a creative way to bring more open space to some of the poorest neighborhoods in Oakland. In the seventies, many homes were abandoned and deteriorating around the city. I talked to one of my duck-hunting friends – the president of Bank of America – and showed him how it could work to everyone's advantage if the bank gave us the dilapidated buildings they owned through foreclosure. Using the land trust construct, we would turn them into gardens, playgrounds, and parks. At the same time, BofA would get a tax advantage for selling to us. It was a classic win-win. Before long, we had acquired a long list of buildings and empty lots from banks and savings and loans that were soon transformed into beautiful public places. For many who lived in the surrounding neighborhoods, these places were their only opportunity to experience nature, a communion I believe every human being has the need and the right to enjoy.

On several of these Oakland parcels, we collaborated with the Black Panther Party, one of many Bay Area movements in the late sixties and early seventies that were rethinking and redefining American society, culture, and values. Today, the Bay Area may be famous for its tech billionaires, but back then, it was ground zero for virtually every counterculture experiment of the time. A few were silly – topless bars and water beds; some were frightening – the Symbionese Liberation Army and the People's Temple; many were joyous – Golden Gate Park Be-Ins and Bill Graham's Fillmore Auditorium, home of the San Francisco Sound of Jefferson Airplane, the Grateful Dead, Sly and the Family Stone, and Santana. But a surprising number were important and transformative, shaping the way we live today.

The *Whole Earth Catalog*, for example, begun in 1968 by visionary Stewart Brand, is credited with weaving together many of the conceptual threads of the period – the back-to-the-land ethic, with its belief in communitarian self-sufficiency; the personal technology revolution emanating out of Stanford and Berkeley; the human potential movement from Big Sur's Esalen Institute; and a new environmental philosophy, which historian Andrew G. Kirk credits me with inventing. In his 2007 book, *Counterculture Green*, he quotes me as saying, "Unless we can make the cities more livable, it is going to be very difficult to preserve natural areas ... city dwellers are really going to be the ones making the decisions on the use of natural resources ... unless some traditional environmentalists moved into the urban areas and helped the cities, the integrity of the total environment would fail."

Stewart generously founded and funded the Point Foundation to pursue an ecotopian and socially just future. He invited me to join the board. Unfortunately, it was impossible to get its unconventional and opinionated members to agree on anything. Some saw political rebellion as the answer. Others believed alternative technology could lead the way. Others, like me and activist Jerry Mander, were wary of technology and believed in protecting the environment through natural means. At his wit's end, Stewart came to me for help in managing the unruly group. I thought of something that worked – a psychiatrist. He facilitated some group therapy for the board members, then arrived at a Solomon-like solution: forget about trying to achieve consensus. Instead, let each board member have a lump sum of money to use as he or she sees fit. Some of the projects they funded were brilliant and groundbreaking. One of them, the Jonah Project, pioneered the whale-saving movement.

With the funds I was allotted, I chose to support a number of worthwhile projects. The ideas of three young men, in particular, struck me as important and far-reaching.

The first, Alan Lithman, accomplished something extraordinary. With three consecutive $5,000 grants, he took a barren, denuded landscape in southern India and brought it back to life.

He showed up at the TPL office one day, explaining that Stewart Brand had sent him my way. He said that after years of working in civil rights and antiwar activism, he began to look inward and headed to India with no particular plan or goal. While there, he discovered Auroville (City of Dawn), a "lab" community that welcomed anyone from any culture who wanted to shift to a new social model, to replace the British colonial system of extracting resources with replenishment of the local environment and economy. As he described the land in and around Auroville, I realized that it was a virtual desert and suggested Alan begin with reforestation. It would be a true act of karma, repaying our debt to the earth by going back to the roots. The challenges were great. The seeds couldn't be planted until the area's impenetrable laterite clay was softened up for digging. The clay couldn't be softened until the aquifer was replenished. The aquifer couldn't be restored until water was brought in from outside. And finally, the soil couldn't support planting without compost. Alan hired local workers to haul tons of water and trash for composting from their villages. They thought he was crazy but were happy for the work. Eventually, Alan's efforts and the earth's forgiving nature paid off. What was once a desert is now a verdant forest supporting more than two million trees. Its recovered habitat is home to the native plants, birds, and creatures who lived there before. After some bumps in the road, the Auroville community of environmental idealists is thriving too, having celebrated its fiftieth anniversary in 2018 with the visit of Indian prime minister Narendra Modi.

According to Alan, who now lives in Eugene, Oregon, "We need to call our own bluff" to prevent the earth's destruction. As for me, it may be the best $15,000 investment I ever made.

My second Point Foundation grantee was Bill Bryan, a bright and delightful guy who was the grandson of the governor of Maine. He had heard about me from mutual acquaintances at the University of

Michigan where he was working on his PhD. Along with Saul Alinsky, I became a subject of his dissertation. We became fast friends and often went elk hunting together in Montana. Bill fell in love with the place, and after working as a community organizer in Hawaii for a while, he moved to Montana permanently. He used his Point Foundation grant to create a management consulting firm for nonprofits. Later, he founded his own nonprofit, the Cook Center for Sustainable Agriculture in the American West. One of its programs, One Montana, became its own organization with the purpose of bridging the urban-rural divide. Bill believes that when you connect people to nature, you connect them to one another and help them realize their interdependency. You get beyond ideological differences and build a lasting constituency for the environment. He admits he's often criticized by more combative single-issue activists, but over the years, his philosophy has served him and Montana well.

I'll always be grateful to Bill for starting a chain of events that led to my establishing an important precedent, one that changed our national parks and public lands for the better. Bill introduced me to Scott Reed, a lawyer whose favorite place on earth was a national forest outside of Coeur d'Alene, Idaho. He was upset that a developer had purchased a piece of private land within the park limits with the intention of building a resort complex. The land in question was an example of an inholding, a legacy from the time the US government first established our national parks and forests. Inholdings are privately owned properties that, for one reason or another, have never been sold by their owners to the government.

Scott was right to be alarmed. Clearly, the proposed resort would permanently deface the wild beauty of the forest. Knowing my work, he hoped I could find a way to prevent the project from moving forward. He had already approached the US Park Service about buying the inholding directly. The park people loved the idea, but they didn't have the funds in their budget to make the purchase. So I made a call to my former employer, The Nature Conservancy, to see if the board would be willing to buy the land and protect it from development until the Park Service could put together enough money to buy it from them. The Park Service folks couldn't guarantee they would buy it, but after meeting with them several times, I was confident they had every intention of doing so.

It seemed like a simple enough proposition, but TNC had never bought land with its own money before. They were used to the role of facilitator – identifying land worth saving, raising the money from donors for its purchase, then handing it over to a land trust. My proposal created great consternation and debate among the board members. About half were willing to try something new and purchase the Coeur D'Alene property, while the other half – corporate, conservative, and antigovernment – were dead set against it. A veteran of TNC board politics, I knew well how to play one side against the other, and the pro-purchase side won the day. TNC negotiated a deal with the developers to purchase the property, the Park Service eventually bought it from TNC, and the national forest remains a wilderness for everyone's enjoyment. Best of all, our work became a model for nonprofit–Park Service partnerships that routinely acquire inholdings in public lands, preserving their beauty and serenity, and saving them from commercial blight.

The third young man I selected for a Point Foundation grant was an aspiring writer named Malcolm Margolin. I'll never forgot Malcolm telling me that he moved his light bulbs from room to room because he could only afford to have a few. The $2,000 grant he received allowed him to write his first book. Just a few years later, he started his own nonprofit book and magazine publishing company, Heyday, with a focus on nature, social justice, and what became Malcolm's lifelong passion – the indigenous cultures of Northern California. Like Alan and Bill, Malcolm represented a very small investment that reaped great rewards for people and the environment.

In 2018, I attended the fifty-year anniversary of the *Whole Earth Catalog's* first issue and found its septuagenarian and octogenarian writers and editors as creative and independent-minded as they were a half century before. Stewart, in particular, is more forward-thinking than ever – much more. He currently heads the Long Now Foundation whose goal for civilization is nothing less ambitious than "fostering responsibility in the framework of the next 10,000 years." To that end, the foundation is sponsoring the construction of a clock intended to run for 10,000 years so, as Stewart explains, it will "do for thinking about time what the photographs of Earth from space have done for thinking about the environment. Such icons reframe the way people think."

Politically, the Bay Area's long-standing live-and-let-live attitude and tolerance for free expression set the stage for the Berkeley Free Speech Movement which, in turn, opened the door for the gay rights revolution and the Black Panther Party's commitment to radical empowerment for the black community.

Then as now, environmentalism was criticized in some quarters as a luxury for privileged white people who didn't need to deal with the more immediate concerns of poverty and discrimination. I've always disagreed with this viewpoint, and founded TPL in the belief that a connection to nature is everyone's right, one that's even more important to the urban poor than to other people.

It turned out the Black Panthers felt the same way.

Portrayed one-dimensionally by government and media as gun-toting anarchists, the Panthers actually believed strongly in the interconnectedness of all people, with one another, and with nature. And they were among the first advocates for environmental justice, especially for the poor who suffer disproportionately from asthma, cancer, and other illnesses caused by the industrial emissions of the factories that are invariably located in their neighborhoods, not to mention the lead and asbestos in their homes.

I worked closely with the brilliant Black Panther activist, Elaine Brown. I was introduced to her by the party's leader, Huey Newton, whom I knew through mutual friends at the Zen Center. When Huey fled to Cuba after being charged with murder, he appointed Elaine to head up the organization.

Elaine told me about her own upbringing devoid of open space and healthy, nutritious food. Using the abandoned and foreclosed lots we got donated from the Bank of America Real Estate division and other sources, Elaine began Gardens in the Ghetto, vegetable plots that provided the nutritious produce for the Panthers' food distribution programs. Soon, we were expanding the concept to neighborhoods in the South Bronx, where Green Guerillas were setting up pop-up gardens in vacant lots.

With Slide Ranch, we took the TPL concept farther afield while still serving an urban population. A supporter of our efforts, Susan Washington

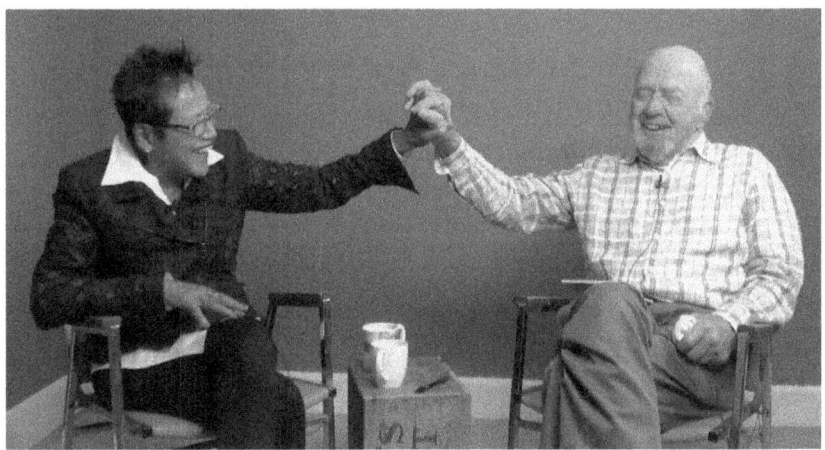

I worked with Elaine Brown, activist and one-time chairwoman of the Black Panther Party, to create community gardens in Oakland.

Smyth, had been wanting to establish a summer camp for poor city kids so they could have a direct and extended experience in nature. At the same time, we learned that Slide Ranch, a beautiful 134-acre coastal property near Point Reyes, was on the market. As it turned out, the only way we could offer the tax savings to win over the seller was to prove that the land would be used to serve the public. I immediately thought of Susie Smyth. Not only was it a win-win for everyone, but it established an important legal precedent for similar transactions in the future.

In a surprisingly short time, the concepts we innovated at TPL proved their worth, and our organization flourished. After five years, we had several million dollars in cash plus eight million in land that was being held for future parks and open spaces.

That's when I made up my mind to leave.

I resigned on TPL's fifth anniversary to become secretary of resources in California governor Jerry Brown's cabinet. Most everyone was stunned by my decision, but it made perfect sense to me. It still does. I had proven my point – that a well-run nonprofit could buy land cheaper than a government agency – and, frankly, I was getting bored. What had started as an exciting challenge had become a daily grind of financial reports and meetings. I knew TPL could run just fine without me. I was eager to move into broader natural resource policy matters, something I could only accomplish in a statewide position.

To this day, I'm asked how I could have left TPL after founding it and building it into such a successful organization. I can explain it best with a story about my friend Don Michael. Don was a University of Michigan emeritus professor and a consultant to American industry. His consulting involved talking to the owners of small companies that had been bought by larger ones. Don would discuss with them the likelihood of their being happy if they stayed on with the companies that had acquired theirs. "Stay on the board," Don said, "and there's about a 50 percent chance you'll be happy. Stay on the executive committee, it drops to 30 percent."

I talked to Don when I was thinking about leaving TPL. I told him that I was bored, that the environmental field was developing rapidly, and I wanted to be part of the change. He went through the same drill with me that he did with his clients. After our talk, he agreed with me and supported my choice. So I said good-bye to TPL then and there.

I think of my work in the environmental movement as something of an art form; it was time to create something new.

The literal and figurative seeds we planted at TPL in the early seventies continue to reap rewards. Community gardens, parks, and playgrounds are still central to TPL's mission, with thousands of acres dedicated to urban open spaces across the country and the world. In 2007, Elaine Brown was the Green Party's candidate for president of the United States, and she currently oversees a cooperative garden in Oakland, owned and run by people who have served time in prison.

Today the Trust for Public Land is one of the largest conservation nonprofits in the world. They have done an excellent job realizing their vision of "land for people," protecting more than three million acres and expanding their scope to include ambitious and creative projects like Chicago's 606 Linear Park and the Climate-Smart Cities Program. (I'm proud to say that even the reluctant Nature Conservancy was eventually converted to the concept of protecting urban open spaces and now includes saving city lands among their myriad programs.)

The executive directors who followed me at TPL were a committed and talented group. Martin Rosen, for example, had a single-minded devotion to the organization that translated into expansive new programs nationwide. Acquisitions during his twenty-five-year tenure include the woods surrounding Walden Pond and the Atlanta neighborhood where Martin Luther King, Jr. was born.

And the delightful Will Rogers, a former beekeeper and honey producer, broadened TPL's vision ever further during his leadership from 1998 to 2018. He initiated TPL's current mission to ensure that every city dweller in the United States has access to a park within a ten-minute walk from home. Will is beloved for his positive and encouraging management style, which brings out the best in everyone he works with.

Still, at TPL's fortieth anniversary party in 2012, I was reminded that I had made the right call thirty-five years before. I'm sorry to say that the entire evening was a spectacle of self-congratulation by TPL's board members. None of the staff – the ones who did all the work – were even invited to the event. And throughout an endless string of pompous speeches, not one speaker acknowledged the exceptional contributions of executive director Will Rogers. He wasn't even invited to speak.

I sat there all evening, growing angrier and angrier, until I finally walked up on stage, grabbed the mike, and gave them a piece of my mind. Then I left the room, disgusted with the whole event..

But that's not what matters in the long run. What matters are stories like the founding of Koshland Park in 1973. It started with a tragedy, a fatal arson fire that destroyed an old San Francisco Victorian mansion converted into apartments. A group of community-minded neighbors saw it an opportunity to establish a one-acre park in their densely populated part of the city, Hayes Valley. About the same time, I got a call from the family of Daniel Koshland, San Francisco business leader, philanthropist, passionate environmentalist, and longtime supporter of my work. They were looking for a meaningful way to celebrate his eightieth birthday. I thought of the property in Hayes Valley and approached the owners, a group of nurses who had purchased the building as an investment. I explained to them how selling to TPL would actually

San Francisco's beautiful Koshland Park rose from the ashes of a tragic fire.

work better for them financially than their other options. They loved the idea. The Koshland family and TPL acquired the property together and gave it to the neighborhood.

But first, we held a surprise party for Daniel, chauffeuring him in a limo to what was still a pile of rubble. When he stepped out of the car, he was greeted by an enormous Happy Birthday banner, a live band, a children's chorus, and hundreds of grateful neighbors. It was the most wonderful celebration imaginable, and a memorable example of how an urban land trust can serve as a catalyst for a mix of strangers – landowners, donors, neighbors, city officials – who all want to do the right thing.

Koshland Park has its own history now, one that spans nearly half a century. It became a model and an inspiration for similar parks across the country and was honored by the American Society of Landscape Architects. Over time, it has changed to reflect the changing needs and interests of the neighborhood, adding a learning garden and a peace wall to its original pathways, hillsides, and playground. But it remains a project of, by, and for the committed people of the Hayes Valley neighborhood.

I think about Daniel Koshland and his park every day when I walk into my office. I have a flagstone table made from a remnant of the old Victorian house that burned to the ground. I had noticed the stone in the midst of the debris at the birthday party, and mentioned it in passing to one of my friends from the San Francisco Zen Center, which is across the street from the park and played a helpful role in its creation. A year later, I was surprised with the beautiful table crafted by a talented artisan at the Zen Center. The transformation of the ravaged stone into something useful and beautiful struck me as very Zen, and very environmental.

I feel great fondness for my time at TPL, and even greater pride that we invented the whole idea of an urban land trust, something that just didn't exist before. My friend and colleague Steve Costa summed up the experience perfectly: "To be able to walk on your work, to go out to those vacant lots and see them become special places … it's a powerful thing."

CHAPTER EIGHT

Lesson Learned | Be a General and a Generalist

Whole Earth Catalog founder Stewart Brand used to call me a "thug for good," a label I consider a high honor. To me, it means that I'm willing to fight for what I believe in, to play hardball for the public good – as long as my actions are legal and ethical. Being a tough guy was rarely necessary at The Nature Conservancy or The Trust for Public Land, where most of my work involved helping wealthy landowners donate their properties. True, I sweetened the deals with tax savings and other financial incentives, but most of the time, our donors were already generous and conservation-minded folks.

Not that I didn't have to do battle from time to time. Like my fight with Gulf & Western over the Marin Headlands. Or taking on the Marin County real estate guys who tried to steal Mrs. Kent's island in Bolinas Lagoon. Or the courtroom confrontation I had with that Montana estate executor who wanted to deny his brothers' children their right to donate the land they had inherited.

But I had no idea how bloody environmental combat could be until I was appointed California secretary of resources in 1978. Taking on the state's water, energy, agriculture, and timber interests – and the politicians who depended on their campaign contributions – turned out to be a challenge like nothing I'd experienced before. And to my surprise, it wasn't just corporations I had to contend with. Colleges and

My one and only government job, secretary of resources for the state of California.

universities, organized religions, nonprofits, and labor unions all fought aggressively – sometimes ruthlessly – for their piece of the pie. I hope the lessons I learned during my five-and-a-half-year tenure will help the next generations of leaders safeguard our state's bounty and beauty for all our citizens, not just those who happen to have money and influence.

Luckily, I was offered the cabinet position just about the time I was itching to leave The Trust for Public Land. It all started when my friend and hero, David Brower, gave my name to Jerry Brown during his first term as governor of California. David was one of the most brilliant and charismatic people I've ever known – a great writer, a world-class mountain climber, a shrewd environmental strategist, and a communicator who brought tears to the eyes of anyone who heard him speak.

The first time I met him was at a cocktail party when I was western regional director of The Nature Conservancy and he was executive director of the Sierra Club. Having increased its membership tenfold, David was the organization's most successful leader since John Muir founded it in 1892.

I walked up to him and introduced myself.

"I'm Huey Johnson. I'm with The Nature Conservancy. I save land."

Brower looked at me and responded, "The Sierra Club will take care of that."

I got it. He saw us as competitors, and he was absolutely right.

As time went on, we remained competitors and became good friends, sharing ideas over lunch whenever possible. We came to admire one another immensely and helped each other out on many occasions. I've always respected David's integrity. In fact, it wasn't long after our first encounter that he walked away from the Sierra Club rather than go along with its endorsement of constructing new nuclear power plants in California. As soon as he left, he cofounded Friends of the Earth, now "the world's largest grassroots environmental network."

One of my fondest memories of David is the morning our paths crossed on Copacabana Beach in Rio de Janeiro. We were both attending the 1992 United Nations Conference on Sustainable Development. I was taking a walk at dawn. There was no one on the shore except a lone figure in the distance walking toward me. As he came closer and grew larger, I realized it was David. We walked and talked for a while. There was something about our sharing an unexpected stroll in that gorgeous exotic setting that has stayed with me all these years.

Back in 1978, a meeting was arranged for Jerry Brown, David Brower, and me to talk over the cabinet job. We met at Green Gulch, the tranquil west Marin County outpost of the San Francisco Zen Center. The spot held special significance for me because a while back I had helped my friend George Wheelwright sell it to the Buddhists on condition it would remain undeveloped. The three of us were joined by Richard Baker, the Center's zentatsu, or Zen master.

Shrewdly setting expectations, David introduced me to the governor: "Huey's a good man but he won't stand in line for anything." Brown offered no reaction, but I knew he respected Brower and would take what he said to heart. The get-together went well, and I got the offer on the spot. As excited as I was about the job, I wasn't willing to accept it unless I was sure I would have both the power to go along with the impressive title and the influence to accomplish important things. I talked with an old friend, Joel Kupperberg, who had served as an aide to the governor of Florida (and who became my replacement at TPL). His advice was a big lesson learned. Joel said it well: "You have to get what you want before you agree to take the position. You can never negotiate after the fact."

So I thought long and hard and came up with four upfront demands: (1) to take a family vacation before I started work, (2) to hire my own deputies, (3) to leave the office every day by five p.m. so I could have dinner with my family (Jerry was famous for staying up to all hours talking with his staff), and (4) to report directly to the governor, not his chief of staff.

These were major demands. The fourth one would be the most difficult for Jerry Brown to accept because it would upset his whole management structure. Yet I knew I would never be able to influence him if my ideas were interpreted by a go-between.

Jerry could tell I wasn't bluffing and would turn down the job if he didn't go along with my requirements. He agreed to all four, adding this advice: "Take care of California's resources, and I'll take care of the politics."

I was off to Sacramento.

Of course, I wasn't the only one who would be moving from Mill Valley to Sacramento. The lives of my wife and two school-age children were about to be totally disrupted. They were proud of me and were good sports about it all, but I wanted them to know how much

I appreciated what they were doing for me. That's why I had insisted on our taking a vacation before I started work and promised them I would be home every night for dinner. And as the cherry on top, I gave my animal-loving daughter, Megan, something she had always dreamed of – her very own horse.

Way back when I was a graduate student at the University of Michigan, I realized that I had no interest in becoming an expert in anything, that I hated the idea of learning more and more about less and less. I preferred to be a generalist, to explore a wide variety of subjects and understand how they were interrelated. Beginning with my travels around the world, I read constantly about natural resources, biology, philosophy, politics, music, history of all types, and biographies of people who had made a difference. The more diversity, the better. When I became a land saver for TNC and TPL, I expanded my reading to include estate law, philanthropy, land trusts, and taxes so I could be an effective administrator and negotiator.

Over time, my extensive reading turned me into a good conversationalist. I've chatted comfortably with a wide assortment of people from all sectors of society including US presidents and CEOs of major companies – even England's Prince Charles and Queen Noor of Jordan.

I'm convinced that being a generalist was the best possible preparation for my new position as secretary of resources. Which is why, as I began the greatest challenge of my life, I returned to the two-thousand-page diary of one of the most accomplished generalists to ever serve our country – Harold LeClair Ickes. Ickes was a newspaperman, lawyer, orator, early crusader for African American rights, environmentalist, participant in the founding of the United Nations, and especially, the brilliant and innovative secretary of the interior during all four Franklin Roosevelt administrations. Since I was about to take on the same job for California's government, it was a stroke of luck that, some years earlier, I had read all three volumes of his frank and detailed work, *The Secret Diary of Harold Ickes*. I sometimes wonder if my choosing to read this vast tome in my mid-twenties was prophetic, or at least planted a seed in my mind. In any case, it now felt like a job manual written just for me. Throughout my tenure, I referred to the book for guidance, just as I did with Aldo Leopold's *A Sand County Almanac* and Juan Ramon Jiminez's *Platero and I*, a beautiful long poem that always inspired me when I needed it most. Another memoir I turned to often was *Cecil Andrus: Politics Western Style*. The four-time governor of Idaho and secretary of the interior under President Jimmy Carter was an ardent environmentalist who fought bravely against special interests out to profit from publicly owned assets. Not long after I arrived in Sacramento, my path crossed with Secretary Andrus's in a most dramatic fashion. More about that later.

Both Ickes and Andrus were tough, effective men with the commitment and stamina to confront constant lobbyist pressures and daily political battles – precisely the gauntlet I was about to face. Not surprisingly, Ickes was labeled a "curmudgeon" and Andrus was known as "irascible." Ickes, in particular, was the target of malicious attacks and investigations, but he loved a fight and accepted conflict as part of his job. And he could give as good as he got in the defense of his beliefs. In the preface to one of the diary volumes, he wrote: "If in these pages, I have hurled an insult at anyone, let it be known that such was my deliberate intent, and I may as well state flatly now that it will be useless and a waste of time to ask me to say I am sorry."

His confrontational style – what I call creative conflict – proved hugely successful. He brought his department out of the Great Depression and contributed to the war effort, all while working to protect public forests, fisheries, minerals, grazing lands, and water.

But even thousands of pages of great advice couldn't prepare me for my first day on the job. When I arrived at the governor's office, one of his assistants walked me over to an ugly fourteen-story office block. He said, "Good luck," and left. That was my orientation.

I entered the building where my fourteen thousand employees worked. The place teemed with people coming and going. I found a directory on the wall and noted the floor and number of my office. When the elevator doors opened on my floor, I was stunned by what I saw. The area was enormous. I would later discover that it encompassed my own large office, several conference rooms, a library, full kitchen, shower, and at least twenty secretaries and assistants at my disposal. (I also had a car and chauffeur at the ready.) In a matter of minutes, I was mobbed by staff needing my approval or signature or instructions or advice. Evidently, nothing could move forward without me.

I quieted everyone down, introduced myself, and explained that I was about to leave for a two-week vacation (one of the nonnegotiable demands Jerry Brown had agreed to). No one knew how to react to this piece of information. I realized I had to leave them with something tangible before I disappeared.

"Read *A Sand County Almanac*. I wouldn't hire anyone who hadn't read it."

Most of them had never heard of the book, but within an hour, every available copy in Sacramento had been purchased, and my staff was on the phone with the publisher ordering many more boxes of Leopold's memoir. It was my first decision in office. More important, I had inadvertently found a way to put my own stamp on the department.

Along with being a generalist, I was becoming a general.

Insisting on a family vacation before I started the job was one of the best ideas I've ever had. Sue and I and our two youngsters, Megan and Tyler, headed off to Europe, picked up bicycles and spent two carefree weeks exploring the continent. Megan, a few years younger than her brother, rode in a seat on my bike, while Tyler did a heroic job of negotiating the crazy traffic of Europe's cities and towns. Before we left California, I promised the kids that if they would stick to our itinerary each day, we would stop at any bakery they wanted along the way. The incentive worked quite well, but there was more to it than that. My children were growing into the kind of people who were willing to do the work it took to learn new things and expand their horizons. I was proud of them for their spirit and determination.

Back from vacation and ready to begin my first real day on the job, I commuted to work as I always had – on my bike. I walked the bike into the lobby, intending to bring it up to my office on the elevator. A security guard approached me and, not knowing I was his new boss, laid into me big time.

"You turn right around and get this thing out of here. There are no bikes allowed in this building. Ever."

I set him straight and took my bike upstairs. As soon as I got there, I made sure the bicycle policy of the building was immediately changed. As a result of the incident, the guard was fired by someone in the chain of command. I had no intention or desire to have the man fired, but as word spread about what had happened, it did cement my authority among the fourteen thousand people who reported to me. And that's not a bad thing.

I do believe that someone has to be in charge. I was determined to earn the respect of those who worked for me, but I didn't need or want their love. That's how I ran the department, and it worked.

Acutely aware of the enormity of my new position, I was determined to be honest with myself about my own strengths and weaknesses. I'm good at seeing the big picture and setting strategic goals, not so good at tactics and implementation. So I surrounded myself with skilled and experienced experts. I brought in top aides whom

I respected and trusted, and had about a dozen administrators who ran their divisions extremely well.

Playing to my strengths, I set out to build an overarching plan that would guide the direction of the department from day one. My battle plan wasn't conceived in a tent on a field of war but rather over a single day in a room at the Sheraton-Palace Hotel in San Francisco. I asked just one person to brainstorm with me because I knew he was all I needed to get the job done. The man was Bill Lyons, a boyhood friend from Michigan who had become interested in politics and served as deputy undersecretary at the US Department of the Interior from 1966 to 1977. Bill agreed and chose to bring along two business professors whom he admired as much as I admired him. One was from the University of Oregon and the other from Stanford Business School.

I'm still proud of what the four of us came up with; today it's mounted on the wall above my desk as a testament to clear, simple, strategic planning. It's as plain a thing as you can imagine. No graphics, no gimmicks. Just a bunch of green index cards with a few hundred words that laid out what each area of my department was responsible for and what we intended to accomplish *as a whole*. The plan came to be known as the Green Card Plan, although it was soon turned into a chart that could be more easily referenced, a guidepost for every employee to use in every situation.

I didn't stew about the plan, then or later on. We had given it our best effort and believed in what we had done. That was enough for me.

A bit about my office. Without consciously realizing it, I furnished it to communicate my beliefs and values so a visitor would know something about me before I uttered a word. My desk was a beautiful slab of redwood burl with a top that was leveled and polished and whose thick sides were left in their natural state. It told people that I valued wildness above all. For seating, I had Mexican *equipales* pigskin chairs plus two canvas chairs that hung from the ceiling. They said I was uncorporate and unconventional. And when I invited someone to sit in one of the hanging chairs, I also learned a lot about them. Whether they were uptight or adventurous, closed-minded or open to new ideas.

Critics stereotyped me, Governor Brown, and his closest advisor, Jacques Barzaghi, as woo-woo hippies who were into Zen Buddhism. Well, if the Zen philosophy of "living well with less, and living close and in touch with the earth" is flaky, I plead guilty. Of the three of us, Jacques probably got the worst rap. Yet I found him to be an intelligent and thoughtful person who cared deeply for the environment and helped me make my case to the governor on many occasions.

With my office decorated and our department plan in place, things got off to a promising start at the Department of Resources. Now it was time to do what I came to Sacramento to do – establish policies that would protect California from the entrenched, well-connected special interests who believed our state belonged to them. Their beliefs were well-founded, thanks to many of my predecessors – I call them ribbon cutters – who found it easier and more pleasant to go along with industry demands. Still, people in similar positions had been successful going to bat for the people – especially my heroes, Harold Ickes and Cyrus Andrus. I was determined to do the same.

Special interests are formidable adversaries, never to be underestimated. They can find vulnerabilities you don't even know you have. I was once asked to speak at a luncheon at the National Press Club in Washington, D.C. In keeping with Governor Brown's theme of a new "age of limits" that would move California toward sustainable energy use and other environmentally friendly policies, I chose to talk about the limits of population. I suggested that California institute policies that would slow growth in our state, like ending subsidies to large families and providing greater support for abortion. I also mentioned the possibility of restricting US immigration.

I had intended to open a dialogue about these serious topics, but they were emotional, hot-button issues of the time, just as they are today. Add to that the fact that everyone in the room was a reporter who would instantly disseminate what I said around the country, and it didn't take long for all hell to break loose. The state legislature was up in arms. Republicans saw me as antigrowth, and Democrats interpreted my remarks as anti-black and anti-Latino.

When I returned to Sacramento, I was greeted by protesters outside my office building. Someone tried to hit me in the face with a placard. Then I discovered that a resolution had been passed by the legislature demanding I resign my cabinet post. It had the overwhelming support of both Democrats and Republicans. To this day, I consider it a badge of honor that I was the only person in state government who could unite California Democrats and Republicans in the same cause.

I was summoned to meet with Gray Davis who was lieutenant governor and acting governor while Jerry Brown was out of town. As I waited outside Davis's office, a couple of my staff members, Kirk Marckwald and Jan Denton, came into the waiting area with serious looks on their faces. Jan was holding an envelope with the return address of the biggest right-winger in the state senate – Joseph B. Montoya. I opened it with trepidation. Inside was a card with a heading that read: "The devil made me do it." It was a practical joke put together by my staff. The three of us laughed our heads off. I appreciated their defusing what had become a pretty tense situation. Fortunately, my meeting with Davis – who didn't like me much either – had no consequences.

The most important thing about the whole brouhaha is how cynically and cleverly the timber, water, and chemical interests exploited one little speech I delivered and turned it to their advantage. They told legislators that they were outraged by what I had said when, in fact, they didn't give a hoot about those issues. Instead, they leveraged a moment of vulnerability to try and get me out of office and replace me with someone more favorable to their agenda.

Thanks to Jerry Brown, their strategy didn't work. He let me keep my job, which I very much appreciated, and the incident soon died down.

Lesson learned: When you agree to be secretary of resources, you have to be willing to quit or be fired for what you believe. Otherwise, don't take the job.

That particular experience with special interests who use underhanded tactics ended happily, but many more of them don't. The most tragic one I know is that of Chuck Stoddard. I intend it as a cautionary tale, a lesson in staying alert at all times.

It happened during the Kennedy administration. I was about to graduate with my master's degree from Utah State and was hoping to land a job at the Bureau of Land Management, a division of the US Department of the Interior that manages hundreds of millions of acres of public lands. What interested me most about working there was the opportunity to serve under its director, Charles Stoddard, someone I had heard great things about. A forester, author, and lifelong fighter for the environment, he spent his time at the BLM working to reduce clear-cutting in western forests and increase grazing fees on public lands. Later, in his native Wisconsin, he oversaw the *Stoddard Report*, documenting the dangerous pollution of Lake Superior. Most important, he was a man of great integrity.

I already had a special interest in working for BLM because of a seminar I had attended in graduate school. The speaker was president of the Utah Wool Growers Association. He was a tall, strong man with cold blue eyes and a face that had seen lots of weather. As is customary, he wore a large brimmed western hat, which he left on his head throughout the event. He drilled us all with a steely glare and stated firmly, "Now you boys may think that the BLM lands where I run my sheep are public. You're fools to think that way. I've had a lot of run-ins about that. I inherited those grazing rights from my dad as he did from his dad, and they are mine. Woe is the bureaucrat who tries to think otherwise."

From that day on, I had it in my mind to get a job at BLM, to take on all those ranchers with their sense of entitlement, and to return public lands to their real owners – the American people. Happily, someone in my graduate program knew Mr. Stoddard, and I was able to get an appointment with him at his office in Washington, D.C. The interview went well, so I was surprised when he told me flat out that he wouldn't hire me. "You don't have any experience dealing with the intense pressure that special interests in resource management put on government employees. You'd be eaten up in a week." As a young idealist, I dismissed his views, but sadly, events soon proved him right.

An idealist himself, Stoddard accepted the appointment as head of BLM, knowing it was dangerous ground for any honorable administrator but determined to clean up the agency. The millions of acres under Bureau of Land Management jurisdiction were mostly dry and degraded terrain abandoned by homesteaders who had moved on. Still, those lands represented millions of dollars in untapped resources – timber, minerals, oil and gas, water – and greedy speculators were always on the lookout for ways to get their hands on those assets.

Because the lion's share of public lands are located in the West, western senators and Congress members receive generous campaign donations from contributors who want access to natural resources. The representatives, in turn, exert tremendous political pressure on federal agencies like BLM.

Chuck Stoddard began his crusade by taking on the timber industry in the Pacific Northwest, which made enormous profits by buying logs off BLM property for a fraction of what they were worth. The BLM's regional director in Portland, Oregon, was supposed to answer to Stoddard, but he was in the pocket of industry, having been left alone by Washington for years on end.

Stoddard decided to transfer the regional director and replace him with someone he knew and trusted, his own assistant director in Washington. But the regional director in Portland refused to leave the job. Stoddard ordered him out immediately. That's when the timber industry, threatened with the loss of millions of dollars, took off the gloves. Senators intervened on behalf of their corporate contributors. Hearings were held. Stoddard remained confident. He was, after all, the head of a federal agency and appointed by the president of the United States.

During this time, I happened to be at a dinner in Missoula, Montana, with Stoddard and Dr. Arnold Bolle, the revered dean of forestry at the University of Montana. Stoddard described his situation, including the fact that a highly respected senator from Montana was in a position to support him but wasn't lifting a finger. A close friend of the senator, Dr. Bolle volunteered to call him and a few other influential people on Stoddard's behalf. Stoddard graciously offered to let Bolle charge the long-distance calls to his credit card.

Weeks later, some attorneys visited Stoddard in his office. They informed him that he had violated the law by loaning out his government credit card and using it as part of a lobbying effort. The lawyers said that unless he resigned, they were prepared to sue his boss, US secretary of the interior Stewart Udall. Stoddard spent several days soul-searching, then resigned.

Stoddard returned to his remote cabin deep in the Wisconsin woods where he lived the rest of his life. While a member of the Wilderness

Society Council, I visited him there. He had gone on to become a respected professor, yet he still blamed himself for what happened at BLM. To me, what matters most is that he went out with his integrity intact. That is a worthy ambition for anyone. Still, an intelligent, accomplished, and moral man was brought down by one thoughtless moment.

Lesson learned: When it comes to high-risk decision-making, never let your guard down.

One more pitfall of interacting with special interests is the corruptibility of science. At one point in my job at Sacramento, an aide informed me he had solid research proving the toxicity of a widely used chemical. He recommended we put regulations in effect to limit its usage. Unfortunately, I was unaware that industries are in the habit of hiring their own scientists – at alluringly high salaries – to manipulate data so it will cast doubt on the most reputable and well-conducted experiments. Despite my having studied science in graduate school – or maybe because of it – I naively assumed that science is science and facts are facts. But at the hearings we had called to make our case, industry chemists found ways to cast doubt on our evidence. Blindsided, we were unable to respond quickly enough to overcome the bad publicity. The outcome: we lost our window of opportunity, and the chemical remained in widespread use.

Years later, I happened to be interviewing an eminent Dutch scientist while on a trip to the Netherlands, With tape rolling, he asked me a question out of the blue. "Why is American science so corrupt?" I was a bit taken aback, then I remembered my experience in Sacramento and realized that he had suffered the same intervention by American corporate interests that I had.

Lesson learned: Science may be objective, but some scientists are not.

Booby traps like these are real and treacherous, yet it's important to remember they're only one side of the story. As secretary of resources, I possessed powerful tools as well. I called them the three Ps – people, position, and public disclosure – and I wasted no time putting them to use. I'm convinced that my tools for protecting the people's resources can be as effective in today's political environment as it was back in the late seventies and early eighties.

Tool #1: People – Staff
By far the most valuable tool at my disposal was my staff. I have the highest respect for the commitment, intelligence, and abilities of nearly all of my fourteen thousand employees. In my opinion, government service is the equivalent of military service. Obviously, government workers don't put their lives on the line or relocate thousands of miles from their families, but they do give up higher salaries and attractive perks they could enjoy in the private sector. Also, people in government service become true experts in fields that improve all our lives. It makes me furious whenever I hear those worn-out complaints about mindless "government workers." Nothing could be further from the truth.

I also found state employees scrupulously honest.

To reward and nurture staff I identified as promising, I began an internship program in my department, something that had never been done in California government before. I got the idea from the Kennedy White House, which had done something similar. I saw the program as a way to work around the California civil service seniority system and give talented employees – especially women and minorities – a chance to realize their potential, move up the ladder, and contribute more to our department. I sent out notices to all of my fourteen thousand employees. There was much excitement about the idea. I received a large number of applications, then selected about twenty or twenty-five people to be part of the first group.

My program paired department heads with interns who were responsible for writing reports and speeches, doing research, and handling whatever else their mentors needed. At the same time, the interns took night school classes leading to a master's degree in planning.

One of the stars of the program was Vera Marcus. When I first met her, she was working as an administrator. Her initiative and brilliance came across immediately. I taught one of her night school courses, and at the end of each class, I assigned readings that would be discussed at the following class. Without fail, Vera showed up having read not only the assignment but the entire book it was excerpted from. On top of that, she had usually done research based on the book's footnotes and citations. Frankly, I almost never knew the answers to her questions. She was just that smart and curious.

Vera's life story was right out of a history book. Growing up in Birmingham, Alabama, in the sixties, she was neighbors with the four little girls whose horrific murder in a church bombing shocked the world. Her sister, the first African American to attend her high school, was often escorted to her classes by state police. When Vera was a senior at the same high school, she was recruited by Princeton University to join their freshman class. Although she had never heard of Princeton when they contacted her, she became the first African American woman to enter Princeton as a freshman and graduate – in three years, no less. A veteran of the civil rights and women's rights movement, Vera was just plain fearless and unwilling to accept defeat. She later attended University of California, Davis Law School and now practices family law in the Bay Area.

After completing her internship, Vera played a key role in my department, advocating for my programs in the state legislature and using her never-say-die philosophy to help me win some historic land protection designations against all odds. More to come on that subject.

Another outstanding graduate of the internship program was Jan Denton. A real Renaissance woman, she had been a nun before working in Sacramento and became an accomplished artist afterward. While serving in my department, she joined Vera in lobbying for our positions in the legislature and did a masterful job.

I could never have performed my job as well – or had as much success – without two extraordinary undersecretaries: Rich Hammond, who served until 1980, and Kirk Marckwald, who was with me from 1980 through the end of the Brown administration in January 1983.

In spite of my insistence that I choose my own aides, Rich was already deputy secretary when I arrived – for three whole days. A Harvard grad, lawyer, past member of the California Coastal Commission, and part of the governor's team dealing with energy issues, he had just started working for my predecessor when she took a new position as president of the California Public Utilities Commission. So in the span of less than a week, poor Rich had been hired to work for a boss who resigned and then met a new boss who immediately went on a two-week vacation, leaving him in charge. Rich assumed he would be fired upon my return, especially after I announced on my first day that I would be letting a few employees go and replacing them to "spice up the stew." I assured Rich that although I couldn't guarantee he would remain my deputy, I would definitely keep him on in some capacity. Once I returned and got to know him, I could tell he was the right person for the job. He understood my belief that we were there for one reason only: to be stewards of California's environment and the public it belongs to. Besides, he related well to politicians and commissioners and almost always found a way to accomplish what I asked him to. No easy task.

After about three years, Richard decided to return to the private sector and recommended someone he knew from the governor's office –

Kirk Marckwald. Kirk was the first person I interviewed for Richard's job and the last. I hired him right away.

A smart, funny, and charming guy, Kirk was a trained professional with a graduate degree in resource management. Like Rich, he possessed a key quality that I lacked – a passion for politics. We had met years before at a conference, and later Kirk told me he felt we were kindred spirits who shared "wild-eyed ideas." I guess that's why he was willing to leave the governor's office and come work for me.

Like Rich and I, Kirk and I became great friends; I even officiated at his wedding. It meant a lot to have someone at my side I could be myself with and blow off steam when necessary. Our way of unwinding was to find an empty office, sip some Applejack, and play cribbage. (I had one other way of dealing with the pressure of the job. I took a short nap or two just about every day, a practice I referred to as "meditating." Somehow my biological clock allowed me to fall asleep immediately and wake up on my own precisely eighteen minutes later. The naps were so refreshing and rejuvenating that I've continued the custom ever since.) Kirk, by the way, went on to found his own successful environmental policy consulting firm, helping corporate clients become responsible environmental citizens. Here's the reason he says he started the business: "After working for you and Jerry, I was unemployable. So I had to create my own job."

It's difficult for me to express how instrumental Rich and Kirk were to achieving my goals as secretary. To begin with, they were my guides to navigating the political realities of state government. And they were masters at finding original and creative ways to achieve seemingly impossible ends. Dedicated and ethical, they rose to every occasion, working tirelessly and cheerfully, often under great duress, freeing up my time and energies to do what I did best. And most of the time, they made it seem like fun.

I talked to Rich and Kirk recently and was gratified to discover that they had picked up a few lessons from me as well.

Kirk said I taught him that there is a wide range of responses to any given situation. If you can rationally assess the correct response, you're already way ahead of your adversaries. Also, most people are conflict-averse and if you're not, you can use that advantage during a negotiation. According to Kirk, I explained it more succinctly: "Punching them in the nose gets their attention." I'll let you decide whether he quoted me accurately or not.

Rich said that, looking back through today's lens, he thought I brought an entrepreneurial spirit to our big bureaucracy and understood the power of inspiring both employees and the public about the environment. He felt I steered the ship based on my values, and that was key to leadership in any organization.

There are a few others I want to call out, Steve Steinhour, Russell Cahill, and Priscilla Grew. Steve has played a part in almost every stage of my career – The Nature Conservancy, Trust for Public Land, Resource Renewal Institute. In all cases, he has used his exceptional legal skills to find new ways to achieve my objectives while ensuring every transaction is legally ironclad. So when the voters passed a multimillion dollar bond measure to buy private property and turn it into public parks, I immediately thought of Steve to oversee the acquisitions. I knew those purchases and expenditures would be under close scrutiny by many factions; I trusted Steve to buy environmentally significant sites and execute the transactions without a flaw. I strongly recommended him to the director of California State Parks, one of the departments under my authority, and he was hired. As I anticipated, Steve performed his difficult and politically sensitive job beautifully. As a result, California's citizens are now enjoying thousands of acres of precious wetlands, beaches, forests, and other open space.

Steve's boss was Russ Cahill, a great guy I first met when he was a park ranger in Hawaii and I was at The Nature Conservancy trying to raise funds to save the Seven Sacred Pools. I didn't know then that Russ's father traced his ancestry back to the Polynesian migration to Hawaii, but his commitment to the land was evident. A decade or so later, I selected him to run California's enormous state park system with its thousands of employees and hundreds of parks, forests, beaches, and monuments. In his book, *Tales from the Parks,* he calls it his "dream job." He ran it like a dream too. To reflect the state's changing population, he brought in more minority employees than ever before. And he found inventive ways to resolve controversial situations. For instance, when faced with the need to remove existing homes from a popular public beach during a housing crisis, he came up with the ingenious idea of dismantling the buildings and reusing the building materials for low-income housing elsewhere. Russ also helped me expand and finalize a project I had begun at The Trust for Public Land – saving the Santa Monica Mountains. There is nothing rarer or more precious in the urban sprawl that is Los Angeles than open space. At TPL, we had managed to save a good amount of acreage in the Santa Monica Mountains for parkland. When I became secretary, Russ and I were able to add significantly to the area. Today, its 75,000 acres are managed by a mix of national, state, regional, and local entities. It is one of the most visited parks in all of California.

Priscilla Grew is the ultimate example of the expertise and professionalism that make government work. Her breadth and depth of knowledge – and willingness to share the glory with others – contributed enormously to my success in Sacramento. A fellow generalist, she excels in taking information from lawyers, economists, geologists, and ecologists and turning it into something far greater than the sum of its parts. While serving as director of the Department of Conservation and a commissioner for the Public Utility Commission, Priscilla's cheerful, calm, and clear way of explaining things helped me understand issues way over my head so I could come to the right decision.

Tool #1: People – The Governor

Around the time Kirk succeeded Rich as undersecretary, we came up with an idea to install solar panels on top of all the government buildings in Sacramento. We were excited by its many advantages: it integrated beautifully with the governor's money-saving philosophy (he was famous for driving an old Plymouth and living in a one-bedroom apartment instead of the Governor's Mansion), it would set an example for California residences and businesses to follow, and it would set the stage for the introduction of tax-advantaged solar legislation.

We enthusiastically presented the plan to the governor. It took just a few seconds for him to shut us down. As we were talking, he did the math and realized that it would take *forty-eight years* to recoup the initial investment and begin saving money. That's what made him such a brilliant politician. He examined every possible angle and consequence of an issue before making a move. I was disappointed by his decision but had learned yet another political lesson that would serve me well in the years to come.

I have nothing but respect and admiration for Jerry. He was a wonderful administrator and human being who never asked me to do anything against my judgment or conscience. I had the same understanding with him; I was ready to resign at any time if my actions brought grief to his office.

With his perfect balance of visionary philosopher and practical politician, Jerry was the manager I aspired to be. It's no wonder he is the only person to have been elected California governor four times. He embodies intelligence, skill, and integrity, and I attribute much of whatever success I had on the job to his confidence and support. I rarely saw him during my time in his cabinet, yet I knew I could always count on him, even in the most difficult times. His usual MO was to simply stay out of things and let me handle it. He may have appeared uninvolved, yet in reality, he knew exactly what was going on and trusted my judgment – something I will always appreciate.

It was unfortunate and unfair that Jerry got saddled with the nickname "Governor Moonbeam." I want to be clear that there was never anything flaky or spacey about him, which his long and accomplished career bears out. (I only wish his presidential campaign had been successful.)

The mix-up came about by accident. The singer Linda Ronstadt, his longtime romantic partner, made an innocent joke to a columnist that had nothing to do with Jerry being a space cadet. It was based on his desire to launch a California space satellite that could be used for communications during an emergency or disaster – something many states adopted in the decades that followed. Soon after the column ran, its writer, Mike Royko from Chicago, said he regretted ever using the word "moonbeam," but almost fifty years later, it's still misunderstood and misused.

There was only one time the governor and I actually parted ways philosophically. Like our mutual friend Stewart Brand, Jerry had faith in the power of technology to solve certain environmental problems. I, on the other hand, always turn to nature for my answers. So when the three of us attended a conference on futuristic space technologies, I spoke out vehemently against what was being proposed. As usual, Jerry was fine with my contradicting him.

Frankly, space exploration has never impressed me that much. I remember that on the day of the first moon landing in 1969, there was a casual mention in the news that the astronauts had used the moon's surface to dump the garbage from their space flight. The formerly pristine surface of the moon was now littered with trash. I found it such a depressing bit of information that I walked down to a nearby river and picked up a single rock – an earth rock. I've kept that rock all these years as a symbol of the importance of concentrating our efforts on protecting this earth, not befouling the rest of the universe.

Tool #1: People – Advocates and Activists

In addition to talented staff members, I found environmentalists a potent tool in winning the battle against the might and money of special interests.

Traditional environmental organizations like the Sierra Club can be valuable allies. Thanks to their help, for example, I was able to get a bill passed requiring timber companies that harvested trees from public lands to plant three trees for every one they cut down. But as often as not, these large nonprofits are more willing to compromise with their adversaries than I am. I understand their "half a loaf is better than none" strategy and occasionally employ it myself, but it isn't the only way to go.

Patty Schifferle comes to mind. As a rafting guide on the Colorado River, she felt fortunate earning her living doing what she loved. But the longer she lived there, the more concerned she became about the imperiled future of wild rivers everywhere. One night around the campfire, her fellow guides encouraged her to follow her dream of moving to Washington, D.C., and doing something significant for the rivers – even if she wasn't sure what that would be. Her comrades scraped together enough money for a one-way ticket, and she was on her way.

Patty's instincts were right on the mark. On her flight to Washington, she struck up a conversation with the man next to her. She explained the reason for her move and happened to mention a few other impressive parts of her bio: she was a Harvard grad and an expert violinist. Her seatmate offered her a job as his children's babysitter, which gave her the income to live in Washington and the free time to become an environmental lobbyist. She developed quite a reputation in

Washington and was later hired as an aide to a California state legislator. So effective was she at California politics that she earned the nickname "Short Cut Schifferle."

I received frequent letters from nature photographer Ansel Adams while I was secretary. His correspondence usually discussed commercial encroachment at Big Sur, where he lived for many years, but he often wrote me to rant about President Ronald Reagan's assault on our public lands. I knew Ansel fairly well and admired him for using his celebrity to fight for environmental causes. He honored me with a gift of one of his gorgeous photographs. I treasure it as well as our friendship.

Some of the most effective environmental advocates I worked with were the last people you would expect – wealthy corporate executives. Why? Because they have a passion for duck hunting. In the rest of the country, membership in an exclusive country club may signal wealth and status, but in California, it's the duck club. In fact, some duck clubs can cost more than a million dollars to join and many thousands in annual dues. Their well-to-do members are willing to pony up still more millions to support nonprofits like the Western Region of Ducks Unlimited and the California Waterfowl Association, organizations that work to preserve migratory duck habitats. They also contribute to the elections and reelections of candidates who will vote according to their interests.

Over time, I earned the trust of these influential CEOs, which made them willing to put in a good word for my side on issues unrelated to duck hunting. Their endorsements often tipped the scales in my favor.

Lesson learned: Advocates aren't always who you think they are.

I've also joined forces with activists – critics might call them extremists – who shared my credo of creative conflict.

The most dramatic and moving example was Mark Dubois. Mark grew up rafting the white waters and exploring the caves of the Stanislaus

```
Mr. Huey  Johnson           Fenruary 6, 1982
Director,
RESOURCES                   (Rt. 1, Box 181
State Capitol               Carmel, Ca   93923)
Sacramento, California

Dear Mr. Johnson,
     BRAVO for you powerful remarks about Watt and the present
Administration.
     We have to take  powerful action to stop these mad men in
Washington! Thank God we live in a country that allows us to
express ourselves! As long as the expression is at the logical
and ideological level we can be as "violent" as we wish. Some
of the new political forces are frightening in their illiberal
directions.  Do you recall the little jingle about "Poor little
Jeep Trail, don't you cry/ You'll be a Highway bye and bye!".The
"poor" little Administration may well lead to a political and
racial disaster - to say nothing of the end of our environ-
mental asperations. I am hoping the Environment will be a basic
issue in the coming elections. But good organization is needed.
And money. We cannot drift any lomger. I wish I were young again;
I would love a good fight!!!       as ever      ANSEL ADAMS
```

> *Photographer Ansel Adams and I exchanged many letters about environmental issues.*

River in Northern California. Later, he helped establish Friends of the River and founded Environmental Travel Companions, a nonprofit that introduced inner-city kids to river rafting and assisted adventurers with disabilities on rafting excursions. So when Mark found out about the impending flooding of the Stanislaus to create the New Melones Reservoir, he was devastated. In a 2006 interview with *The Union Democrat,* he described his reaction "like a mother rescuing her child from being hit by traffic." He chained himself to a rock in an area of the river that was about to be flooded and refused to tell anyone where he was. After about a week, he was tracked down and removed from his spot. I supported Mark's action, as did the governor, and we used the authority of the state to take on the federal government and try to prevent the flooding. The battle went on for about a year, but eventually the reservoir was filled, destroying the life cycle of salmon and steelhead trout and erasing a uniquely beautiful natural place.

Yet, in the long run, this turned out to be a positive story, a case of losing the battle and winning the war. Mark's courageous act – which he calls "heart politics" – became national news, educated the general public, and signaled the close of the era of large dam construction in America.

A word of caution: activists don't always work for your cause. One of the most famous *anti*-environmentalists during my tenure was a conservative Republican named B. T. Collins. A Vietnam War veteran who had lost an arm and leg in battle, B. T. had been appointed director of the California Conservation Corps by Governor Brown. As you would imagine, he and I held diametrically opposed opinions when it came to the environment, especially regarding the use of pesticides. To promote his point of view, B. T. came up with an unusual stunt. He plopped himself down on the capital steps and drank a beaker full of malathion, a pesticide whose use I had recently restricted. It was a risky stunt, to say the least, but he survived, and the story got tons of publicity. In spite of his going on to hold several other prestigious positions in state government as well as serving in the state assembly, B. T. Collins is still best known for downing his malathion cocktail.

Tool #1: People – Secret Friends

I found my adversaries to be valuable assets too. The lobbyists and advocates who represented various special interests were experts in their fields. I figured, "Why not take the opportunity to learn what they know and use it to my advantage?" I invited them over regularly to pick their brains and get their point of view. This practice had the added benefit of building friendly relationships that could help me in any number of ways. For example, lobbyists often unwittingly revealed their strategies to me, allowing me to beat them at their own game. Or they sometimes did me a favor like letting a bill lapse past its intended legislative session or putting it on the bottom of a stack of documents to be signed.

During these meetings, I took the opportunity to gently make my environmental case to my rivals. Occasionally, I would win one of them over on a particular issue.

That's when they became my secret friends.

A stunning example was the head of Dow Chemical Company for the western states. We often argued in the press because I was in charge of signing off on any chemical he wanted to market in California. I rarely obliged.

On one occasion, he showed up at my office and asked if he could close the door. Suddenly he confessed, "I'm on your side. My family and I are very environmental. Too many of the chemicals we sell are dangerous. We haven't tested them. I just wanted you to know."

He described how he frequently advocated for a shift in environmental policy at Dow board meetings, but his nemesis, the vice president of agricultural chemical sales, would invariably interrupt with the same diatribe: "Dow gets a third of its profits from chemicals. Do you want to lose a third of your profits?" Then the VP would cite the board's fiduciary responsibility to its stockholders. So nothing ever changed.

He was eager to help and asked me how he could. I suggested he resign from the California Agricultural Aircraft Association, one of my toughest rivals. Not only were they spraying with Dow chemicals, they were leaving the chemical barrels in the fields to pollute nearby fields and streams, kill fish, and endanger the health of agricultural workers. He promptly resigned from the association, resulting in a significant reduction in pollution.

I had another secret friend, this one in the timber industry. His bosses were particularly hard-nosed, so he would do me a favor once in a while by "accidentally" forgetting about one bill or another that he had been instructed to push through the legislature.

One more type of secret friend was a lobbyist who held a grudge against a fellow lobbyist, perhaps someone who had received a special favor from a legislator. To get back at this perceived injustice, he or she would often give me some juicy information or pass it on to a reporter as an unnamed source.

Sometimes secret friends feel too much pressure to act on your behalf, even if they are sympathetic. During one of our get-togethers, I asked the largest single recipient of public water in California if he would join me in opposing the greedy irrigation interests who were demanding far more water than they could ever use.

"This will surprise you. I'm on your side. I love fishing, for instance, and have a custom-made fly rod I always carry in my suitcase. I have enough money so my family and I will never need any more. But I'm loyal to my friends, and *they* don't believe they have enough water.

We all know the government will crack down on us sometime soon, but at this point I have to say no. Sorry."

Yet even reluctant secret friends may eventually find the conscience to covertly help the environment. So it makes sense to maintain a pleasant ongoing relationship with every last one of them.

Tool #2: The Power of the Position – Integrity

Mark Twain is supposed to have said that if you tell the truth, you don't have to remember anything. Whether he said it or not, it's the truth. Every position comes with political baggage, established before you ever get there. The wise and correct thing to do is be honest, straight, and narrow in carrying out the responsibilities of the job.

That I spent more than five years in state government and left with my integrity intact was possible only because I valued being ethical above all else. Frankly, I never found it that difficult. The laws were on the books, and I simply followed their instructions. It didn't take long for the word to get out around Sacramento that I was incorruptible, so no one bothered to tempt me with money, gifts, or a job after I left office. I knew that if I ever gave in to tempting offers, I would become instantly vulnerable to blackmail and public disclosure. I preferred a clear conscience and sleeping well at night.

I saw many heartbreaking examples of officials who got in over their heads. Some had drinking problems or got involved with women other than their wives. Obviously, the lobbyists had them wrapped around their fingers. But sometimes the pressures were more subtle – a small request to liberalize a regulation, postpone a ruling, or perform a simple favor – and before an official realized it, he had compromised his integrity.

I had one simple rule that always kept me on track: the public owns the state's resources and I am their employee. Just as a physician takes a Hippocratic oath and an attorney pledges to uphold the law, a government employee has the obligation to uphold the public trust.

Tool #2: The Power of the Position – Being Proactive with Special Interests
After all my criticism of special interests, it may seem surprising that I considered them a helpful tool in performing my job. The truth is, they are an essential part of the democratic process and have as much right to fight for their causes as anyone else – if they do it legally and ethically. So if you don't figure out how to deal with them, you can never achieve anything of consequence in a government position.

The first thing to know is that natural resources are a fixed pie, which means stakeholders are forever competing for the same assets. If resources are nurtured and protected, they continue to provide value generation after generation. Unfortunately, most industries choose to take what they can get in the moment without concern for sustainability. Moreover, these companies have a sense of entitlement. It is based on a long history of our government giving away millions of acres of publicly owned lands to private interests in the form of land rushes, oil wells, mineral mines, forest timber, grazing fields, and more. With a population that continues to grow every year, the resources pie must be cut into ever smaller pieces in order for everyone to have clean air, fresh water, and plentiful food – and beaches, forests, and mountains to enjoy.

Federal agricultural subsidies and price guarantees are another part of the power dynamic. Even with the explosion in technology and other fields, agriculture is still California's largest industry. There seems to be little logic in how subsidies are doled out, with precedent and influence determining the allocations.

To maintain the status quo, special interests have permanent offices and large staff in Sacramento to woo elected official to their side. But it is usually more Machiavellian than that. Sometimes a lobbyist who represents one industry will back the bill of an unrelated industry, knowing the favor will be returned when needed. Or a lobbyist may have multiple favors "in the bank" that they can trade with second and third parties to ensure support downstream. Thus the system is convoluted and loaded with potential problems for an innocent political appointee suddenly put in charge of millions of dollars in public funds.

One effective tactic used by special interests is to convince politicians to "starve" oversight committees. A few key committee members in their pocket is all it takes. Water rights interests like the wine industry are notorious for this. California remains one of the few states that have never confirmed ownership of water in every stream. It is estimated that at least a third of water claims are actually false, allowing undeserving private interests to use water they don't own.

In spite of all these inequities and the overarching influence of money, I can honestly say that the majority of the lobbyists I knew were honest and ethical and knew more about their subject matter than anyone else around. Which is why I sought them out and met with them frequently. I regularly set up debates between well-informed lobbyists and members of my staff as a way of testing our knowledge and refining our arguments. Some lobbyists, like those I got to know in the fishing and water industries, collaborated closely with my team, hammering out solutions that promoted their businesses while maintaining high environmental standards.

They were my staunch opponents, to be sure, but I knew I had to make certain concessions to them or I would never get anything accomplished. In their hearts, most of them wanted to do the right thing. They understood better than anyone that the future of their industries depended on the sustainability of our groundwater, streams, rivers, and oceans. So when I provided them with the opportunity to be good guys, they often took it.

Obviously, it was important to establish relationships with the people everyone was trying to influence – senators and assembly people. Vera Marcus and Jan Denton, the legislative assistants who had graduated from my internship program, did a great job cultivating friendships in the legislature, so when we needed to get votes, they were both known and trusted around the capital.

Being proactive also meant maintaining cordial relationships with everyone. Just because you're friendly and respectful doesn't mean you have to compromise your principles. When we spent the day doing battle in the press, we made it a point to socialize in the evening or have breakfast the next morning with our adversaries. I rarely enjoyed those evenings out, but I knew it was the only way to keep things civil in order to accomplish my goals. One exception: the Refuse Collectors' parties. Their hors d'oeuvres were spectacular.

A sense of humor helps too. After I delivered a scorching speech about the anti-environmental practices of the cattle industry, a small article appeared in the California Cattlemen's Association newsletter:

> We never thought we would see the day when State Secretary of the Resources Agency, Huey Johnson would be on the CCA membership team.... Lo and behold, we have received 2 written statements and several phone calls from cattlemen who said they hadn't paid their 1981 CCA dues yet, but if we were as bad as Huey said we were, they were going to do so right away.

The fact is, few legislators want to be known as enemies of the environment.

The closest friend I had in the state capital was Peter Behr, a Republican senator. At the same time, I had a very powerful friend in the California Senate and later, the US Congress – Phil Burton.

The best way I know how to sum up my relationships in Sacramento is with a story I once heard about Harry Truman and Herbert Hoover. Although they were fierce political rivals with opposing ideologies, Truman described Hoover as "my friend and I am his." The more I got to know my opponents in Sacramento, the more I understood what Truman meant.

Tool #2: The Power of the Position – Final Checkpoint

Soon after I became secretary, I discovered the immense power of the pen. Whether I signed or didn't sign a document determined if a project would move forward or not. During my time in office, there were perhaps eight thousand major construction projects, dams, releases of new manufacturing substances, and industrial chemicals under my authority. I wasn't capricious about it; still, I used my influence many times to stop something in its tracks that was not in the interest of California's public trust.

I followed the recommendations of my well-informed deputies for most of the projects, but I handled the most environmentally and politically sensitive ones myself. As you'd expect, when the time came for my yea or nay, I experienced tremendous political pressure. I got endless calls and visits from lobbyists and elected officials urging me to sign off on a project. They always used the same rationale: the projects meant growth and jobs for their district. What they never said was that it represented rewards for a representative's financial donors. I can honestly say that I always made my decisions based on the intrinsic value of the project, not on politics.

One example of this endless ritual: a small group of lobbyists stopped by my office to urge me to approve a project that would benefit them. They were nice people, as lobbyists almost always are. They didn't dislike me; they just saw me as a barrier they needed to get beyond.

They walked in and seemed nervous as they attempted to seat themselves in the unusual chairs of my huge office. After a few moments of small talk, they got to the point. "We all know the tradition is that someone in your position compromises to meet the needs of your clients, but you seem unwilling to compromise." I was astonished that they considered themselves my clients, rather than the people of California. Sadly, that was what they had learned to expect.

Then I spoke. "Well, I wouldn't rule it out but I've sworn to protect the public and our state's environment. The laws on the books are my guidelines, and that is what I follow. Besides, you fellows have managed to develop this tradition of compromising, and as far as I can assess, it has been to the detriment of the voters and the environment." I continued, "It may have happened a little bit at a time, but it's had a notable impact on things like water and air pollution, pesticides, and our forests. So during my time on the job, if there is any compromising to be done, you are going to do it for a change."

They said, "You know you won't be popular with that attitude."

I thanked them and they left. But, in the end, they compromised.

To quote Harry Truman again: "The buck stops here."

Tool #2: The Power of the Position – Policy

As much as I talk about the pressure private industry places on government, it can work the other way too. By setting policy as a government official, I was able to give environmentally friendly businesses valuable support.

For instance, I provided an entrepreneurial guy named Alvin Duskin the use of five thousand acres of government land at Altamont Pass to launch the world's first wind energy farm. I knew Alvin a little from San Francisco where he was an energetic activist, fighting battles for high-rise height limits, which he lost, and turning the island of Alcatraz into a park, which he won. He had become well versed in energy issues working as an aide in the US Senate and came to me to see if I would guarantee a price that would give his wind energy idea a chance to succeed. I ran it past the government attorneys who said they didn't really know whether it was okay for me to help him or not. I interpreted their response as, "Sure, go ahead," then wrote a letter to Alvin agreeing to a price. He showed my letter to some banks who lent him the money he needed, and his project took off. Altamont produces clean energy to this day, and California continues to lead the nation in renewable energy. Unfortunately, the original turbines killed raptors and other birds, but they are now being replaced by newer, less harmful models.

Tool #2: The Power of the Position – The Law

When all else failed, I knew I could resort to a highly effective tool most officials never think of – California state law. Surprisingly, there are many laws on the books that are rarely fought for or put into effect because of custom or the influence of private interests. I decided to change all that by enlisting the services of the state's capable attorneys, all experts in California laws, statutes, and regulations.

On a number of occasions, I used our lawyers to sue the federal government, particularly when the jurisdiction of the state and the nation were at odds. Many lands in California are legally owned by the US Forest Service, but as secretary of resources for the state, I was sworn to protect them for the people and environmental health of the state.

Federal lands have been exploited by private interests for so long that those interests have come to see those regions as their own (like the sheep rancher who spoke to my class at Utah State). In just about every case, they lease lands for far less than market rates and buy natural resources for a fraction of their worth. The ultimate loser is the US taxpayer who pays the bills for building roads and maintaining the property.

Such was the case with the head of the US Forest Service in California. He was willing to give the timber industry anything it wanted without regard to its effect on the health and sustainability of the forests and the rivers and streams nearby. I started by disagreeing with him in public and wound up suing his department. We actually stopped the logging for a couple of years, but ultimately timber interests and labor unions had too much influence in Congress and got what they wanted.

The feds owned so much California land that, half the time, they didn't know what they were fighting for. There was one pristine area in the Sierras – the Golden Trout Wilderness – that was so remote, almost no one had ever set foot there. (This was in the days before GPS and satellite photography.) A federal wilderness bill was in the works. I sued the Forest Service to make sure the area would be left in its wild state and not turned over to private companies. Happily, I won this one, and an unspoiled area remains wild today.

I took on the federal government yet again when migrating ducks were being poisoned by agricultural pesticides north of California, then dying when they arrived in our state. The timing was too urgent to take the feds to court, so I ordered the immediate cancellation of California's duck-hunting season. Everyone was angry with me – duck hunters, farmers and ranchers who rented out duck blinds, and the US Forest Service whose authority I defied. But it was the right thing to do, and eventually everyone quieted down and I survived the episode.

The whole affair reminds me of something I used to tell my deputies and aides in the Department of Resources. We're not here to win a popularity contest. We're here to be respected.

Tool #3: The Power of Public Disclosure
Never underestimate the importance of getting the word out. I learned that lesson long before I went to Sacramento, first at my sales job, then

at TNC and TPL. As a public official, the impact of what I said increased exponentially. On the one hand, every word I uttered was scrutinized by the press. On the other hand, millions of people paid attention to whatever I had to say.

In my experience, once the public knows the whole story, they generally want to do the right thing. So I made a point of speaking regularly at community and citizens' groups both large and small. I found I could be persuasive enough to convince a neutral crowd of my point of view and tough enough to face down a hostile crowd, often winning them over.

Engaging with the public is exactly how I won a big victory for Mono Lake. Just east of Yosemite, it is all that is left of a once vast inland sea. Few life-forms can survive in its saline waters, but one that can is a shrimp favored by migrating birds. Because of this abundance of food, the western gull flies there to nest. A stream that flows into the lake contributes just enough fresh water to prevent the lake from becoming too saline for aquatic life to survive. And around that time, a group of environmentalists had reintroduced some rainbow trout into the stream. They seemed to be doing well. Nevertheless, water interests decided they needed to divert the fresh water from the stream to Southern California. If they were to win, it would devastate the shrimp, trout, migratory birds, nesting gulls, and the entire future of Mono Lake. The Audubon Society and other environmental groups had filed a lawsuit that was dragging on and on.

Frankly, I didn't see how my department could help. Los Angeles water interests held so much sway in state government, there was virtually no chance the environmental lawsuit would win. But at the last minute, my staff and I came up with an idea. We scheduled a series of public hearings around the state, hoping that several weeks of press coverage would bring enough pressure on politicians to save the lake. To further educate the public, we decided to publish a report and summary of all the proceedings.

Many of the hearings were held in rural locations in blazing summer heat. Sticking out like a sore thumb was the team of formally dressed lawyers who represented the LA water interests. They said little and took copious notes.

One of them came up to me at our session in Bakersfield. He leaned over my desk and placed his hands in fists on the table. His face was dripping with sweat from the hundred-degree temperature.

"Damn, I wish I was on your side on this one." Then he turned and walked away. That was the first moment I felt a twinge of optimism.

When it came time for the judge to issue a decision, he ruled in favor of the environmentalists, quoting from the summary of my department's report. Did the fact that our report came from a state agency give him the impression that it was somehow a law? It occurred to me that it might, but I sure wasn't going to ask him. What is clear is that the force of public disclosure allowed us to win against all odds. Best of all, the stream is still flowing into Lake Mono.

Another time, I was able to win over a small and far more hostile community in a backwoods timber town of Northern California. As I entered the high school gym to give my talk, I noticed that everyone had been given an ax handle to pound on the floor whenever I said something they disliked. They pounded many times, but I didn't let it scare me off as it had public officials in the past. Even a dartboard with my face on it – put together by timber interests – didn't stop me. By the time I was finished talking, the audience members realized that I wasn't their enemy. In the end, the lumber companies harvested scientifically in accordance with the law, and the forest remained healthy and sustainable.

Sometimes just the *threat* of public disclosure is enough to accomplish your goal. That's what happened when a turf war developed between two University of California campuses. UC Santa Cruz, the newest in the UC system, was focused on the environmental sciences and had established a program in organic farming. The acknowledged world leader in agricultural sciences, UC Davis saw the new program as competition. As I looked further into the matter, I began to see what was really going on. Davis received large donations from the pesticide and chemical industries who had lobbied the legislature to have the Santa Cruz organic farming program eliminated from the budget. I talked to the governor, and he agreed to have the money put back in the budget, but the funds were removed yet again. After this back and forth occurred three times, I called the chancellor of the University of California and asked him to come to my office to explain why his budget was being tampered with. When he showed up, I told him point blank that I was about to start a public relations blitz disclosing what was going on, especially the influence of lobbyists on UC budgetary decisions. We talked frankly, and he admitted that there was some rough politics swirling around the university system. I reassured him that the UCSC environmental program was concerned with developing environmental generalists and systems thinkers, not the scientific specialists who were trained at Davis. I explained that the two curricula actually complemented each other and benefited the UC system as a whole. Before long, he agreed to fund the Santa Cruz program and assured me that this time the budget would remain. And it did.

I see this incident as an example of my comfort with calculated risk. I hoped the chancellor would agree with me, but if he didn't, I was fully prepared to carry out my threat. After all, I had right on my side. If the whole story came to light, I had handled myself honestly and with integrity. Only the other side would have been in trouble. That's the value of a clear conscience.

To sum up my one and only experience in government, I am proud that I was able to accomplish most of what I set out to do for the environment and the people of California. In Chapter Nine, I'll describe what I consider the three biggest achievements of my time in Sacramento. Before I get into that, here are a few other things that give me great satisfaction.

I helped put an end to the growth of nuclear power in California.
When I took the job, I already had an aversion to nuclear energy, which was under my jurisdiction as secretary of resources. My feelings stemmed partly from an emotional experience I had with my son. He was still young enough to sleep in a crib when I found him crying one night. I asked him what was wrong, and he said, "Nuclear bombs." I have no idea how he had heard of atomic weapons at such a tender age, but I was moved by the impact it had on him. On a more practical level, no one had ever convinced me that nuclear waste could be safely stored.

As I took office, there was already great pressure to approve the construction of Sundesert, a new nuclear power plant proposed for Southern California near the Arizona border. The momentum to build came from several factors: a few years before, the nation had been plunged into a gas and economic crisis caused by the OPEC (Organization of the Oil Exporting Countries) oil embargo; $100 million had already been spent on the project; and San Diego Gas and Electric had invested millions in a PR campaign to win over the public. To win me over as well, the head of San Diego Gas and Electric took me on a private tour of Sundesert where he proudly told me something that would have been funny if it hadn't been so frightening. "If we ever have an accident at Sundesert, there's absolutely nothing to worry about. The prevailing winds will send the radiation to Arizona or Nevada. California will be just fine!"

It felt very lonely at times, fighting for what I knew was right. But eventually, with the governor's support, the tide began to turn in my favor. I received an unannounced visit from the president of San Diego Gas and Electric. He walked into my office and surrendered. "I'm giving up. You win." And that was the end of Sundesert.

Stopping Sundesert in its tracks was a great victory, but California still operated nuclear power plants around the state. Laws regulating the facilities had been passed, but the legislature hadn't funded them adequately to ensure the public safety.

Then Three Mile Island happened and everything changed. The worst nuclear accident in American history terrified everyone and presented me with an opportunity to get the legislature to make California nuclear plants safe and secure. I knew that the attention span of politicians and the public was short, and I had to act fast. I immediately established a task force made up of people from twenty-five state agencies. Their assignment was to prepare a comprehensive report with recommendations for the legislature to enact and departments to implement. I needed a real star to oversee the immense operation. I remembered a whip-smart young man, Phil Greenberg, who, a few years back, had volunteered with my wife to pass Proposition 15, a ballot initiative favoring greater restrictions on nuclear power plants in California. Phil was an interesting guy. He had played guitar in an English rock bank and served as an aide to a congressman. What I liked most about him was how he had built a portable ramp for his old dog Fella so he could go up and down the stairs without suffering.

When I got in touch with Phil, he was a graduate student at UC Berkeley. I didn't let that get in my way. I contacted the professor who was Phil's advisor and somehow convinced him to let Phil reduce his workload at Cal and come to Sacramento to oversee the task force. So eager was I to get Phil on the job immediately, my wife and I invited him and Fella to live in our garage so he wouldn't waste any time looking for an apartment.

I put a lot of pressure on Phil to get the nuclear report done. I'm sure he dreamed often of returning to Berkeley and focusing on his PhD program, but he stayed with the challenge he had agreed to. He once described me as "an inspiration and a tough boss." I'll leave it to you to read between the lines of that cryptic description.

Phil jumped in with both feet, supervising the research, holding hearings and press conferences, thoroughly inspecting nuclear facilities and even hospital X-ray equipment, all while enduring tremendous political pressure from legislators and private energy interests. In a startlingly short amount of time, he presented his findings and recommendations for the inspection and regulation of the plants that could be made safe and the decommission of the ones that couldn't. He also provided detailed methods for safely storing nuclear waste, then topped it all off with the approval and backing of the American Medical Association and the American Dental Association. Thanks to Phil's superhuman effort and the effective management of my right-hand person, Rich Hammond, California became a national leader in nuclear energy safety. And it's all still in place today.

I thoroughly enjoyed working with Phil. He had a way of making the toughest times fun and full of laughs. We don't put those qualities on our résumés, but work would be a lot more enjoyable if we did.

I allocated state funds to plant trees in urban areas.
Based on my experience with TPL, I used money from the general fund to set up programs in cities around the state. Local citizens were delighted to maintain the trees, and they have transformed neighborhoods in Oakland, Los Angeles, and many other cities.

I carved a track in stone.
In other words, I made sure the practices I established would continue

after I left. I think this is critically important to success in government since everything can disappear with the election of a new administration.

I helped save almost fifty thousand acres of open space in my spare time.
After the citizens of California allocated funds to buy significant privately owned land for public use, one of my department aides suggested I check out a wetlands area just a short drive from the office. Evidently, the owner allowed duck hunting on his land, so I decided to look over the property and get some hunting in at the same time. I had no idea I was about to have one of the most moving and memorable experiences of my life.

It turned out the land was located along the Cosumnes, the only undammed river in the Central Valley. It is surrounded by the last remnants of the riparian oak forest, savanna, and wetlands that once covered the valley. The area is also home to a stretch of the Pacific Flyway and a seasonal refuge for the magnificent sandhill crane.

I knew almost none of this as I settled in and started scanning the skies for the migrating ducks I hoped would be my dinner that evening. It was pouring rain, my favorite kind of hunting weather. For some reason, I find great pleasure in being out in a storm, my body dry under my rain gear but my face streaming with water. But this time, instead of finding ducks, I saw enormous gray birds – sandhill cranes, I would find out later – flying toward me, and then, astonishingly, landing right next to where I sat. Their neutral-colored body feathers were set off by brilliant red-orange forehead plumage, Keeping as still as possible, I watched their elegant movements and hoped they would stay forever. They remained long enough for me to begin a lifelong romance with cranes of all kinds. Everything about them fascinates and inspires me: their physical beauty, exuberant dances, sociable behavior, lifelong monogamy, and miraculous migratory flights.

I've since learned that I'm not alone in my passion for cranes. For millennia in cultures around the world, they have been the subject of myth, symbol, and art. Depending on the culture, they represent peace, joy, happiness, honor, vigilance, and love. In Japan, where they are said to live for a thousand years, cranes are a symbol of hope and have become the most iconic of all origami paper sculptures.

Cranes as a symbol of hope took on even greater meaning for me as I followed the resurgence of the nearly extinct North American whooping crane from twenty-three birds in 1941 to more than eight hundred today. I became a board member of the International Crane Foundation, and when I founded my own organization, I knew exactly what my "mascot" would be. My office displays lovely artwork of every kind – paintings, drawings, etchings, sculpture, photography – that has been given to me by artists and friends who are aware of my lifelong passion. And every holiday season, friends of Resource Renewal Institute receive a small crane gift of their own to thank them for their commitment to nature's renewal.

After my extraordinary first encounter with the sandhill cranes, I became obsessed with providing a lasting refuge for them along the Cosumnes River. At first, I thought it would be a perfect acquisition for the state of California, funded by the recent bond initiative. I tracked down the landowner and discovered that he, like many farmers and ranchers I've met, had a hate on for "the government" and would never sell to the state. In fact, he threatened to cut down every last one of the rare valley oak trees on his property if he were "forced" to turn it over to the state. On the other hand, he was quite happy to make a deal with a private buyer.

I contacted my old friends at The Nature Conservancy, hoping they would step in to protect the sandhill crane habitat. Unfortunately, they didn't have the funds. I was still hell-bent on creating a Cosumnes

Preserve but couldn't find a way to make it happen. Finally, I came up with a plan. I decided to mortgage my own home in Mill Valley and buy the land myself. I took a gamble that TNC would eventually find the money and buy me out. It was a risky move, but there was no way I was going to let my cranes face an uncertain future. After about two years, TNC did provide a solution: instead of buying the Cosumnes property from me, they swapped it for a different parcel of land that I kept for a while and eventually sold.

TNC has been an excellent steward of their acquisition, expanding the area extensively through public-private partnerships and agreements. Today, the Cosumnes River Preserve has grown into a vast forty-six thousand acres, some used for agriculture, the rest for parkland, water recreation, and trails. It is now registered with the National Natural Landmark Program as an illustration of the natural history of the area.

As urbanization encroaches from Sacramento in the north to Stockton in the south, I'm forever thankful I had the foresight – and the nerve – to take out that mortgage so many years ago.

The year I owned the land along the river was my last year on the job. My family was ready to move back to Mill Valley, so I bought a trailer and lived alone on the Cosumnes property during the week. I felt a little like Huckleberry Finn on his raft, and like Huck, I found it a lovely way to live. I would catch a fish or bag a duck for dinner. My neighbors were black-tailed deer, otters, beavers and bald eagles, not to mention my beloved sandhill cranes. Living there reminded me why I had chosen to become secretary of resources to begin with, and it gave me the resolve to face the politics and bureaucracy every day.

If the trailer was my refuge at night, I had another way to reconnect with nature during my lunch hour. Along the Sacramento River, not far from my office, a thoughtful person had hung a rope that would allow a passerby to swing over the riverbank and jump into the water, providing instant relief from Sacramento's stifling summer weather. I took advantage of it often, so I wasn't that surprised when my picture showed up in the newspaper – thoroughly enjoying one of the California resources I had sworn to protect.

CHAPTER NINE

Lesson Learned | Think Big. Act Bigger.

In 1978, I left the land-saving business for state government so I could make sweeping changes to California's environmental policy for generations to come. To my way of thinking, I accomplished three things that were far-reaching and comprehensive enough to fit that bill – Investing for Prosperity, RARE II, and preserving twelve hundred miles of wild and scenic California rivers. These three victories have something else in common. They were all considered lost causes at the time.

Investing for Prosperity
I'm pretty sure the idea for Investing for Prosperity (IFP) had been percolating in my mind for a long time without my realizing it. As a confirmed generalist, I had spent my whole life noticing and exploring the connectedness of what others consider unrelated. For example, the Green Card Plan – the one Bill Lyons and I devised for organizing and managing the California Department of Resources – integrated the entire organization for the first time and got fourteen thousand people working toward a common goal.

IFP, which literally came to me one night in a dream, was another exercise in creative generalism. To this day, I call it my dream project: partly because it began as a dream, partly because it meant so much to me, and partly because it was so damn difficult to make a reality.

The story of how IFP became California law is also a good lesson in the political process – the push and pull, the give and take, the expected and unexpected ways that people with power behave. Most important, it is an example of how fresh thinking and inventive proposals can bring the least likely people over to your side.

When I woke from my incredible dream, I could barely wait until six thirty a.m. to call my department heads and tell them to meet me at a coffee shop for a seven a.m. meeting. They arrived sleepy and confused, but I was burning with energy and excitement. I explained to them my new overarching strategy for the future of California's natural resources. At the time, I called it the One Hundred Years Plan. It had two basic components: the first, to responsibly manage our state's precious resources so they could thrive for centuries to come; the second, to make sure both the state general fund and the industries that profited from our resources contributed financially to replenishing what they used. For instance, oil companies with offshore drilling leases should be obligated to support sustainable energy alternatives. Best of all, the funding would no longer be fought over every year in the legislature. Instead, the state would make long-term investments in its natural resources and energy conservation, saving money in the process. The idea was simple, which is always good, and it integrated all of California's resources as well as those under federal jurisdiction within our state – forests, wildlands, fish and wildlife, water, soils, the Pacific Coast, public parks, minerals, and energy.

My team and I brainstormed all day long, and by five p.m. we had a complete plan of action. We brought in our large staff of graphic designers and artists and tasked them with visualizing our work in a clear and compelling way. They labored long into the night, and by the next morning, the presentation was ready to show to Governor Brown.

What's that old saying about the best laid plans? The very day we completed our beautiful presentation, the governor announced he was cutting all department budgets by 10 percent because of Prop 13, the 1978 voter initiative that slashed state property taxes. The news meant our entire program would be dead in the water before the governor had even heard about it.

I don't know what came over me, but I grabbed the presentation and ran to Jerry Brown's office. Usually, I was quite respectful of his office protocol, but this time I flew past his assistant's desk and barged right in. He was in the midst of a meeting with his senior financial advisors. I interrupted their conversation and told them I had a new idea that was far too important to let die. The future of California was at stake. I furiously flipped through my charts and graphs, supporting my emotional rant with rational evidence.

Then I heard myself say something that surprised me as much as it did the others in the room. Maybe more.

"For the idea to succeed, we need to do more than just replace the 10 percent budget cut. We need another 10 percent on top of that."

Jerry looked slightly amused. He turned to his lead advisor, Richard Silberman, and asked him what he thought.

"I think it's a hell of a good idea. We can find the money somewhere."

And that's how Investing for Prosperity and the Department of Resources got 20 percent more funding than any other department or agency in the state of California.

It's also how I learned one of the biggest lessons of my life: Nothing is ever cast in stone.

Whatever Jerry had originally mandated, whatever his advisors had told him to do, regardless of whatever the other cabinet member might say, he changed his mind on the spot and gave me what I needed.

I returned to my office and shared the unexpected good news with my staff. There was whooping and hollering all around, but after a few minutes, the new reality set in. What were we going to do with our 10 percent windfall? The truth was, our whole plan had been based on the original budget. I knew the additional funds could do a world of good, I just didn't know how. It was a classic case of "Be careful what you wish for."

So I reached out to my department heads for more ideas. I told them to think about what they had always dreamed of achieving but never had the budget to implement. Before long, we had expanded our plan, creating perhaps the most comprehensive, integrated, and sweeping program ever proposed for environmental sustainability. It became the cornerstone of my work in Sacramento, resulting in a billion-dollar investment of public and private funds dedicated to California's natural resources. It also set the direction for the rest of my career, especially the international Green Plan movement I started after leaving government service.

Even with Jerry Brown's approval and our extra funding, IFP faced one more enormous hurdle – the California legislature. Only the legislators had the power to make IFP a law. With its maze of political parties, special interests, committees, and subcommittees, the legislature presented us with a gauntlet of challenges. I remember approaching a group of Republican legislators about the plan, many of whom held tremendous power based on their decades of seniority and committee chairmanships. They literally laughed in my face. I knew they would, of course. This was the same crowd that had introduced several bills to eliminate the Department of Resources altogether.

So I began to think of ways I could use my own political power. After all, if I could bring the governor on board in a matter of minutes, IFP was clearly a powerful and convincing idea.

First, I won the endorsement of respected and influential mainstream organizations like the League of Women Voters. Next, I reached out to broad-based environmental groups like the Audubon Society and the Sierra Club. Although they often advocated for single-issue causes like endangered species preservation and wilderness protection, they immediately saw the value in IFP's statewide integrated approach. The labor unions were next. Once I explained to them how IFP would lead to job growth, they became supporters too.

Then I thought of another option. What if I went directly to the politicians' financial contributors? The future viability of their industries depended on an ongoing supply of timber, clean water, healthy soil, and plentiful fish. The same was true for the banks and other businesses that needed thriving industries to support the overall economy. And as I've said before, you never know who might turn out to be your friend, secret or otherwise.

The uncle of one of my staff members was the head of IBM's West Coast operation, so I started with him. Sitting in his office, I explained how IFP would benefit the state economy and give his thousands of California employees a better quality of life. He was easily convinced, so I kept going.

"Would you mind putting your lobbyists to work on behalf of IFP?

"Help yourself."

Next I went to Bank of America. I had a good track record with them based on my work at The Trust for Public Land. They had seen firsthand how big business and environmentalists could work together to everyone's advantage.

With BofA and IBM in place, it became easier to convince other big companies to get on board, if only so they would look like good guys too. Southern Pacific joined up, along with many others.

Naturally, the industries that would be the most affected by IFP were the most skeptical. The timber industry is a good example. For

My boss, Governor Jerry Brown, signing the Resource Investment Fund into law.

them, we devised a program that showed understanding of their situation and addressed their needs. California timber production had been declining for decades due to air pollution and lack of stewardship. Smaller logging companies – about a third of the industry – were causing most of the problems because they couldn't afford to be both environmentally responsible and pay for things like road and bridge repair. Our solution was to create a grant program that would fund whatever improvements they asked for *as long as they also put a forest renovation stewardship program in place*. The timber industry quickly signed on.

With so many of their corporate donors and constituents crossing over to our side, the Republican legislators began to see things differently. Soon Investing for Prosperity became California law.

Over the years, IFP's integrated, collaborative approach to managing both resources and government agencies has been wildly successful. The innovative sustainability programs we initiated forty years ago have saved California billions in dollars, acre-feet of water, and kilowatts of electricity. Over the decades – even when less environmentally minded administrations were in power – the program has become institutionalized and ingrained in the government ethos and culture. IFP proved that government, so often disparaged as wasteful and unimaginative, could be a leader in efficiency, productivity, and the achievement of our most important shared goals. Today, it is as much a part of California life as the mountains, rivers, and forests it was created to protect. It does me proud.

RARE II

Sometimes we face a decision so immense and high risk that it changes us. RARE II was one of those decisions for me. I made the choice to take on the president of the United States, the Congress, the most powerful politician in California, even my friends and colleagues in the Sierra Club, Wilderness Society, and the National Resources Defense Council in order to save more than two million acres of wild public lands.

RARE II stands for the Second Roadless Area Review and Evaluation. As you would imagine, it followed something now known as RARE I. The original RARE was a report mandated by the Wilderness Act of 1964 to evaluate some fifty-five million acres of public lands across the nation and determine how much of it should be included in the National Wilderness Preservation System. Finally published in 1973, its recommendation was to protect about a fifth of the reviewed land and open the rest to commercial use. Controversial from the minute it was released, the report led inevitably to a second review, with industry and environmentalists locking horns once again. This time, the amount of public land at stake was larger than it was during RARE I. President Carter had agreed to an advisor's recommendation that RARE II settle the status of *all* unclassified public lands – more than sixty-seven million acres – and put an end to decades of squabbling.

As a young teacher in Idaho, I had campaigned for passage of the 1964 Wilderness Act, the one that created RARE I. I got my junior high school students involved in the fight – to the chagrin of many of their politically conservative parents. It was my first exposure to the environmental opportunities and pitfalls surrounding the vast federal holdings in western states where ranching, mining, and timber interests believe not just in less government but in no government at all (except for the below-market federal leases they enjoy). Eventually, most Idahoans did see the value in the legislation and were happy it passed.

Although I had already become secretary of resources when RARE II was under review, I rarely heard a peep about its progress, and it fell off my radar. After a while, I found out why. All those involved in the process – the Forest Service, politicians, businesses, *and* environmental organizations – suspected I would never agree to what they were cooking up, so they deliberately kept me out of the loop. They were absolutely right. The agreement allowed for about 2.3 million acres of pristine California wilderness to be opened up for "mixed use" – everything from ski resorts to deforestation to mining.

Why would California's major environmental organizations agree to this plunder? Frankly, they had been worn down by years of negotiations with Phillip Burton, the tough-as-nails, all-powerful California congressman who had absolutely no interest in environmentalism or land preservation. They felt lucky just to get an occasional brief audience with him, let alone have any chance of winning him over to their point of view. So it's understandable that they adopted the "half a loaf" philosophy so common in politics. Get what you can and call it a day. I agree there is a time and place for give and take, but this wasn't it. The entire character of California was at stake.

To be fair, Burton had pushed back against big timber and other industries during the negotiations, amassing far more wilderness in the agreement than what industry had originally wanted. But he did so primarily to assert his power over California Republicans during a critical reapportionment year.

I only became aware of this unholy arrangement about a month before it would have become a done deal. I learned about it through phone calls from "secret friends," employees of the Forest Service who shared my commitment to preserving wildlands. Knowing my reputation for taking on the federal government and special interests, they sought me out. They were infuriated by something called computer stuffing. They were being ordered to produce and submit official maps of wilderness areas no one had ever seen or stepped foot on. Their superiors were using the unsubstantiated data to create formulas that justified their predetermined conclusions. It was a scandalous practice, pure and simple.

As the incriminating calls kept coming in, I got more and more agitated. "You know, to hell with this. I'll risk it all and sue the bastards and stop this whole process." Nevertheless, I kept my thoughts to myself, knowing that a lawsuit would unleash a firestorm of criticism and trouble. To be honest, I couldn't think of one person who would be on my side.

Before long, word leaked out to some of the mainstream environmental advocates who had negotiated the agreement that I was against what they had signed on for. They asked to meet with me in my office. Many were longtime friends. "Huey, we've been negotiating for years, and we've cut a deal. True, we've given away a lot, but what we've got is okay. We think you'd be better off just doing it."

I told them about the shenanigans that had been going on at the Forest Service.

"A lot of these lands nobody's seen, been on, or evaluated. They'll be logged and mined, and we will never know what we lost."

"Look. We've cut a deal, and we don't think it can be improved upon. Don't sue!"

As they left my office, I looked at them and thought, "They've let their egos get in the way of their responsibilities."

After everybody in my department had gone home, I stayed at my desk. Dusk came on, and it started to rain. The rain was pounding on the windows. I sat there alone, thinking about my choices. "I can go along with the agreement, be one of the boys, save some wilderness and enjoy congratulations all around. Or I can say, 'No, this is just plain wrong. We'll lose a lot of wilderness we shouldn't lose.'"

I went for a long walk in the rain, feeling lonely as hell. I came back to the office, sat some more, and thought some more. I stared at the stack of legal documents I had asked a few young staff attorneys to prepare. "I'm going to do it. I'm going to challenge them."

I called the young attorneys at home and told them I was ready to move forward with the first lawsuit to ever challenge the Forest Service on how it established wilderness designation areas. My staff lawyers seemed to be in favor of what I was about to do, and it felt good to have at least a few people who agreed with my decision.

The staff attorneys had a simple and compelling legal argument for putting the brakes on the RARE II agreement. Astonishingly, no environmental impact study had been performed or filed on the tens of millions of acres under review. The study was required by law so we decided to make the case that the agreement was invalid.

We kept our plans close to the vest because even my environmentally minded boss, Jerry Brown, would not have agreed to take on this lost cause in the making. It would have been far too costly to him in political capital and relationships.

We could bypass the governor, but there was still one major obstacle standing in our way – Attorney General George "The Iron Duke" Deukmejian. His signature was required to file suit with the US District Court in Sacramento. A conservative Republican with strong ties to the resources industries, he sometimes tried to do favors for his corporate friends, favors I would invariably kill when they reached my desk. Clearly, he was not one of my fans and would never agree to what I was about to do. He did have a sense of humor, though. Whenever we passed each other in the halls of government, he would say, "Good morning, you son of a bitch," and I would respond, "Hello, you bastard."

Putting our heads together, my wily staff and I came up with a high-risk approach to solving the Deukmejian problem. One of our team knew someone whose girlfriend, an ardent environmentalist, worked in the attorney general's office. She was familiar with his habit of arriving to work early in the morning and signing the stack of papers waiting on his desk. Our "plant" knew that he typically reviewed the first five or ten documents before signing them. After that, facing a busy day of meetings ahead, he would grow impatient and sign the rest without looking them over. So on this particular day, our friend placed our documents at the very bottom of the pile. It worked. Deukmejian signed off on our lawsuit without reading it, and we proceeded to file our attorney-general-approved documents in the district court.

Once they were filed, our documents became available to the public, and it wasn't long before all hell broke loose. Only the governor stayed cool, waiting to see how the whole affair would play out.

True to form, Phillip Burton went ballistic. Although I had cultivated relationships with many California politicians, I had not met Burton, purposely avoiding him because of his terrifying reputation. When he got word about the filing, he called me and introduced himself in the following way: "Mr. Secretary, what is this horseshit that you're going to file suit?" He went on from there. "I've been working for two years with these green freak friends of yours. We worked out a whole deal that can't possibly be improved upon, and now you're going to kill it with a lawsuit? Tell me you're not going to do this, because it's stupid!"

I had already gone through an exacting decision-making process and was confident I had made the right choice. No one – not even Phil Burton – was going to turn me around now. "Congressman, I'm doing it. I filed the suit already."

He went into a rage. Finally, he said, "You son of a bitch, you better win or else," and slammed the phone in my ear.

Next, I got a call from Democratic congressman John Seiberling of Ohio, who was as different from Phil Burton as you can imagine. A polite, quiet man, he had played key roles in the Israeli-Egypt Peace Treaty and the Richard Nixon impeachment hearings. He contacted me in his capacity as chairman of the House Committee on Public Lands. He calmly made the same case against the lawsuit that Burton had, and I calmly refused to change my position.

I continued getting pushback from people all over the United States. Someone would call from Illinois or Maine or Alaska, always with the same story: we spent two years in negotiations. All sides are happy with the outcome. You've thrown a wrench into everything. You're throwing out the baby with the bathwater.

In response, I called a meeting at my office for all the complainers from around the country. A large number showed up. First I let them vent, then I let them have it. I told them they should be ashamed of themselves, that they had sold their integrity down the river along with America's patrimony.

Nobody said a word. They looked stunned as they left the room. I never received another call from anyone about the lawsuit.

As the months went by, I tried to focus on my other responsibilities, but the lawsuit was always on my mind. After about six months,

I got a second call from Phil Burton. His tone of voice had changed dramatically. It was soft and friendly.

"Mr. Secretary, I'm calling to tell you that you won. Congratulations and thanks for the lesson. When can we get together for lunch?

I still can't believe it was Phil Burton, of all people, who gave me the momentous news: thanks to my lawsuit, more than a million acres of the original 2.34 million slotted for development would be disallowed and sent back to Congress for review. In a seventy-two-page landmark decision, Judge Lawrence Karlton was pointed in his criticism, vindicating my assessment that the Forest Service had pushed through its recommendations without complying with the Environmental Policy Act. In his inspired and witty decision, the judge noted that the Forest Service had described some of the lands in question as "a mountain" or "a river." He added, "One can hypothesize how the Grand Canyon might be rated: Canyon with river, little vegetation." And in a directive I had never asked for or expected, he expanded the designated areas to include not just California but parts of Nevada and Oregon too.

To save face, the Forest Service appealed the ruling to the Ninth District Court of Appeals, a higher court with wider jurisdiction. They lost again. The final ruling and the publicity surrounding it had a more far-reaching outcome than the original. Over the years and decades that followed, environmentalists from across the country would use our precedent to set aside millions of additional acres as wilderness.

Phil's congratulatory phone call meant we had a lot more work ahead of us – convincing Congress to protect the more than one million acres the judge had opened for reevaluation. This time, however, I had one of Washington's most powerful politicians on my side. Phil's experience with RARE II had turned him into a confirmed and zealous environmentalist. He even gave me a T-shirt to commemorate our

> **"When you deal with exploiters, you have to terrorize the bastards."**
> **Phil Burton 1926 · 1984**

One of Congressman Phillip Burton's more colorful quotes.

shared victory. Only he could have come up with its memorable phrasing: *"When you deal with exploiters, you have to terrorize the bastards."* It is still on my wall.

To celebrate our success, he invited me to lunch, asking me to pick out a place I enjoyed. I chose Greens, a well-known restaurant then owned and run by my friends at the San Francisco Zen Center. It is often described as the finest vegetarian dining in the country.

A steak and potatoes guy, he took one look at the menu and blurted out, "What the hell is this?" A damn vegetarian place? I glared at him. "Just shut up and order something." He laughed and said okay. He had the soup, loved it, then went on to order several other dishes. "Damn. This is good." After that, we had lunch together many times, always at Greens. To me, the story says a lot about Phil as a person – opinionated but open to change.

After RARE II, Phil became a loyal ally. For the rest of my time in office, he always took my calls and showed his trust in my judgment by doing what I asked him to do. He worked to block timberland sales. He convinced chemical companies not to use substances known to be poisonous. He prevented people from stealing water they didn't own. And although his interventions didn't always work and the usual lawsuits continued to fly back and forth, he saved my department a great deal of time and effort, freeing us up to accomplish much that we wouldn't have be able to do without his help.

Phil and I remained friends for the rest of his life which, sadly, ended shortly after our time working together. He was just fifty-six when he died. It pleases me that today he is remembered primarily for his environmental accomplishments. In addition to RARE II, he played a key role in establishing the Point Reyes National Seashore and the Golden Gate National Recreation Area. A few years after his death, an area of Point Reyes was named the Phillip Burton Wilderness. Fittingly, it is the only designated roadless wilderness on the California coast.

I've talked about how expertly Jerry Brown balanced his political role as governor with his strong commitment to the environment. He often kept his distance from controversial issues, letting me take the lead – and sometimes the flak. He did just that during the course of RARE II, but once my victory was secured, he wrote me this generous letter:

Huey D. Johnson…

Few people in public office ever have the opportunity to make decisions which can significantly affect the quality of life of not just our own, but of future generations. Your perception of the importance of the RARE II issue at a time when few others shared that understanding created one of those rare opportunities. Your commitment to maintaining and improving the quality of our natural resources provided you with the courage to make the proper choice.

Your decision to challenge the U.S. Forest Service in court over its conduct of RARE II and its anti-wilderness recommendations for designating the future uses of the last six million acres of roadless areas in California National Forests has resulted in court decisions that not only confirm your judgment but provide important standards against which any such future studies must be measured.

For my part, I was proud and thrilled by what RARE II had achieved. I think what I said to the press at the time captures how I felt then and still feel now:

> I am confident that one hundred years from now it will be viewed not as the victory of one agency over another, or even as one state over the federal government. Instead, it will be viewed as a landmark victory for all the people of California and the nation.

Of course, it doesn't take long for daily life to take over, and for years, I barely gave RARE II another thought. Then, several years ago, I was looking through a box of old personal papers and accidentally came across the *Congressional Record* describing what had happened all those decades before. Phil Burton had kindly autographed it and sent it to me. I never read it back then, just hurriedly filed it away. Reading it for the first time, I discovered that far more acreage had been saved then I was originally aware of. It suddenly hit me what an immense accomplishment RARE II had been – for the environment, to be sure, but also for me personally. I had risked everything by defying everyone I knew – friends, colleagues, superiors, politicians, and enemies who couldn't wait for me to fail. It was all on the line – my job, my reputation, my personal integrity. Yet this was why I had taken the job in the first place. So I embraced the challenge and let the chips fall. In hindsight, I'm convinced that pressing forward, regardless of the consequences, gave me the courage to make other tough choices that came my way for the rest of my life.

Wild and Scenic Rivers

Wild rivers are a linchpin of the natural environment, providing food and habitat for native fish, birds, mammals, and plants. They replenish the soil of riparian forests, prevent erosion, and support healthy ecosystems near and far from the rivers themselves. Wild rivers also support local economies by attracting visitors who love to hike, fish, hunt, raft, and camp out. Unfortunately, Southern California is a desert with a large population that demands our rivers be dammed and channeled to provide their cities and farms with water. So we have very few rivers and streams that remain free-flowing.

Back in the day when you could be both a Republican and an environmentalist, there was a state senator named Peter Behr who was both. He authored a bill preventing the damming of twelve hundred miles of wild California rivers located in the northern coastal redwood forests.

Senator Behr's bill was voted into law, but as often happens, nothing could really come of it because there was no funding allocated for its implementation. On top of that, timber, agriculture, and water industries initiated a barrage of lawsuits and other actions that would keep the law from taking effect far into the future. Another slick move by special interests was inserting a requirement into the legislation that both the California secretary of resources and the US secretary of the interior had to sign off on it or it wasn't valid – something that almost never happened in the controversial world of water rights management. All in all, there were enough stumbling blocks to ensure that Behr's legislation would remain a law in name only. Politicians like this kind of thing because it lets them have it both ways: they can tout their environmentalism to the voters while assuring their corporate campaign donors that nothing will change.

The year was 1980, and I'd had a couple of years on the job to figure things out. I was confident we would find one way or another to put some teeth into the law and safeguard the North Coast wild rivers. But it was also an election year, and as the Iranian hostage crisis dragged on, it became increasingly clear that Ronald Reagan would be elected our next president. If we waited much longer, there was no way Reagan's future secretary of the interior would agree to protect the rivers in question. (That appointee turned out to be James Watts, the most anti-environmental person to hold the office up to that time. Maybe ever.)

We had to move fast, and we did, putting together all the documents necessary to present to the federal court, and if it ruled in our favor, to the current US secretary of the interior, Cecil Andrus. Based on the court's past rulings and the questions asked from the bench, we had a good feeling we would win. Another good sign: Secretary Andrus had taken steps to protect a million acres of Alaska wilderness before Reagan took over. We were optimistic he would help us too.

Then, with the holidays upon us and only a few weeks left until Reagan's inauguration, we were suddenly stopped in our tracks. We discovered that someone in my department was working for the other side. He had purposely withheld critical information from us that only he would know: the lawsuits filed against us by industry required an environmental impact study, a process that can take months or years to complete.

We were toast. There was no way we could meet the deadline now. But as I've learned time and again, if you refuse to accept a cause as lost, something or someone will appear to save the day. This time, it was Vera Marcus, the brilliant young woman who had been the shining light of my internship program and become a gifted legislative lobbyist for my department. She rose from her seat and announced she would take it on. She seemed to have no doubt whatsoever that this herculean task could be accomplished. She set about putting together a team of tireless young lawyers and law students who were willing to work day and night.

They were joined by the San Francisco social justice law firm of Garry, Dreyfus, McTernan and Brotsky – known for their defense of the Black Panthers – who took up our cause and charged us nothing at all.

Many others got on board, selflessly canceling their holiday plans to join us in a building I rented in Sacramento because all the state facilities were closed till after New Year's. I remember one wonderful young man in particular, Jonas Minton, who had been married just a few weeks. His wife, who hadn't seen him since our work began, called him on Christmas morning and pleaded with him to come home for their first Christmas dinner as a married couple. He made it home for dinner, then rushed back to our makeshift offices for another all-nighter.

At the same time, Vera organized the requisite public hearings along the North Coast. Some of the notices appeared only a few hours before the hearings were scheduled to take place. Understandably, our opponents were up in arms, yet I knew we were following the letter, if not the spirit, of the law.

Thanks to Vera's leadership and her never-say-die team, we finished up the report. All our work was completed before Reagan's inauguration, with a few days to spare.

But the race to the finish line didn't end there.

We still didn't have a ruling from the federal court. What's more, President Carter had directed his cabinet officials to resign on the day before the inauguration, then join him for a special farewell dinner that evening at the White House. What if the court ruled in our favor but Secretary Andrus had already resigned? All our work would have been for nothing.

So we asked Cecil Andrus if he would be kind enough to postpone his resignation in case we were lucky enough to get a ruling in our favor that day. At four o'clock on the very last day of the Carter administration, the federal court ruled in our favor. Two of our lawyers who had holed up in the courthouse awaiting a decision grabbed the necessary documents and ran over to the White House. Secretary Andrus was already at the president's dinner when he was informed that my department lawyers were waiting outside the door. He excused himself, sat down, and authorized our documents, placing twelve hundred miles of the Eel, Klamath, Trinity, Smith, and lower American Rivers under the federal Wild and Scenic Rivers System, thereby preventing any future dams, water projects, or development. And just for good measure, he included the forest areas that surround the rivers – preserving the beauty of the terrain, protecting the riverbanks from erosion, and, best of all, infuriating the timber companies even more than they would be already.

After weeks of stress, anxiety, and exhaustion, I felt exhilaration, joy, and gratitude. It was a great day for all of us.

My undersecretary at the time, Kirk Marckwald, says that the experience taught him to look at a situation from every possible angle. Vera, who had spent her life on the front lines of the struggles for civil rights, women's equality, and gay rights, says she is more proud of her work saving our wild rivers than anything else she has done. She sees environmentalism as the "mother movement," without which all other causes are impossible. "We all have to find a way to make the earth last," Vera told me in an interview. "I'm pretty proud of being part of the wild rivers movement. I have to say it's the most exciting historic thing I did. And what's wrong with that?"

There is a postscript to this story, one that could have resulted in a financial disaster for me personally. I've mentioned before how people in public office must never let their guard down, or their opponents will jump in and ruin their lives. I had committed one faux pas when I first became secretary by speaking out for population control. Since then, I had kept my powder dry. However, with the whirlwind of activity surrounding our attempt to save the rivers, I had made a grievous mistake. It is illegal for a state employee to spend public funds on a project without first getting legislative approval. With everyone gone for the holidays and all the state buildings closed, I had spent state funds renting a private facility and preparing all the necessary documents – without getting authorization. My opponents in the timber and water industries, incensed over losing the rivers case at the last minute, decided to sue me personally for the unauthorized expenses. The amount was $332,000, which would have put me in debt for the rest of my life. The plaintiffs formed a front organization called the National Outdoor Coalition and elicited the help of Don Rogers, a Republican assemblyman from Bakersfield who had a vested interest in keeping the water flowing to Southern California. The matter got a great deal of press and publicity. We went to court, where I was represented by a quality San Francisco law firm – Howard, Rice, Nemerovski, Canady, Robertson & Falk. Fortunately, we prevailed and the plaintiffs chose not to appeal.

Luck had stepped in and rescued me from financial ruin. Come to think of it, everything about saving the rivers involved a lot of luck. I was lucky to find out about the missing environmental impact report while there was still time to act. Lucky to have a superwoman like Vera Marcus take up the gauntlet. Lucky that the federal court ruled in our favor on the last hour of the last day. And lucky Secretary Andrus left his dinner party to sign his name.

Now, decades later, I'm taking the time to look back at Investing for Prosperity, RARE II, and wild and scenic rivers, and to think about what it all meant. Why did I care so much about wilderness? Why did I put my career at risk time and again? My answer is that, for me, it comes down to things like bears, pine martins, all kinds of quail and grouse, deer and elk, trout and ouzels – and making sure they have somewhere to be. I'm hoping that now, and a hundred years from now, someone will find strength from my example. Strength to make the effort. Strength to take the risk.

CHAPTER TEN

Lesson Learned | Living the Lessons

Do our lives have themes and motifs like novels and symphonies do? I think so. The difference is, writers and composers figure out their themes ahead of time while, in life, we only discover our patterns by looking back over the years.

Now, with the luxury of hindsight, I've come to believe the theme of my life's work has always been to find bigger and better ways to protect the environment.

I can see now that even when I didn't realize it, I was constantly pushing the boundaries of my work so it would have greater impact. For instance, whenever I was working to save a piece of land for The Nature Conservancy, I invariably tried to acquire the properties adjacent to it. That's how I was able to enlarge the area I preserved in Maui from the original Seven Sacred Pools acquisition to a ten-mile swath that climbs eight thousand feet from the Pacific Ocean to the Haleakala Crater. In the same way, I now think I founded The Trust for Public Land as an expansion of the whole concept of land saving to include the urban landscape. And clearly, my choice to leave TPL to become secretary of resources was based on my need to create not just a community garden here or a playground there – important as those are – but to have a lasting impact on the natural resources of the most populous and influential state in the nation.

That same need to accomplish more led me to start the Resource Renewal Institute (RRI), a small nonprofit think tank I founded in 1985 and still work at every day. With RRI as my sandbox and a staff of dedicated, creative, and cheerful environmentalists by my side, I've been able to pursue what we have come to call "innovations for a sustainable future." As climate change ratchets up the challenges we face, RRI seeks out solutions that are practical, affordable, and, especially, applicable worldwide. Personally, RRI has allowed me to stay put in the place I love, the Bay Area, and pursue any and all the ideas I find worthwhile. Lucky me.

Truth be told, phrases like "pushing the boundaries" and "think tank" would never have crossed your mind if you had bumped into me in the early part of 1983. Jerry Brown's second term as governor had just ended, and he was replaced by my old nemesis, George Deukmejian. As a member of the Brown cabinet, I was out too. On one day, I had a billion-dollar budget, fourteen thousand employees, and a car and driver at my beck and call. On the next day, I was a guy without a job or a plan for the future.

To be fair, I intentionally put myself in that position. I probably could have remained in state government in some capacity or set off on a political career, but I had no interest in either of those prospects. The Nature Conservancy and The Trust for Public Land offered me their executive director jobs, but again, I wasn't excited by land saving anymore, and both organizations had grown way too establishment for my taste. I knew I was ready for something new, for a different way to be an environmentalist. I just didn't know what that was. So as I had done when I left graduate school, then corporate America, then TNC and TPL, I allowed myself to live with uncertainty, trusting that I would eventually find my way. In my opinion, this is one of the most valuable lessons I can pass on to others. If you find yourself out of a job and can live on a shoestring for a time, make the most of it. Try to relax and reflect on where you've been and where you could go. Think for a while. Stop thinking for a while. Make use of the unique perspective that comes from stepping away from what you're used to doing.

My own step away took the form of a funky office at Fort Cronkhite, a former army barracks in the Marin Headlands. A complex of old white wooden buildings, the fort is set in one of the loveliest places imaginable – a small cove and beach along the Pacific Ocean with a view of the Golden Gate Bridge in one direction and a lagoon dotted with shorebirds in the other. Depending on the season, I would see migrating whales, leaping dolphins, or brown pelicans dive-bombing for fish. On some of the surrounding hills, you can still find old World War II batteries and fortifications where the military searched for enemy ships and planes. I had a particular attachment to Fort Cronkhite because it was part of the area slated to become Marincello, the enormous residential development I successfully fought against when I was western regional director of The Nature Conservancy. Today, it's all part of the Golden Gate National Recreation Area, preserved along with thousands of other coastal acres from San Mateo to Point Reyes.

For some months, I simply relished the peace and solitude, riding my bike each morning from nearby Mill Valley, then settling in for a day of reading and reflecting and enjoying the beautiful setting. It was the ideal antidote to five-and-a-half high-pressure years in Sacramento. Sure, I had enjoyed the perks of being a cabinet officer, but now that it was all over, I really didn't miss them at all. I was happy to return to a life of simplicity, to drive a used car until it fell apart and buy my clothes at secondhand stores – especially if it meant I could go my own way without compromise.

Before long, some wonderful friends approached me with a generous offer to support my period of reflection, believing that something worthwhile would come of it. They reminded me of those Renaissance patrons – the Medicis come to mind – who used their wealth to fund the work of artists, scientists, and thinkers. I was and am deeply appreciative of their unusually kind support. To honor them, I named my yet-to-be-defined venture The New Renaissance Center.

My benefactors – all of whom had backed my efforts in the past – trusted me wholeheartedly. They knew from experience that I would spend their money carefully. No fancy offices or bloated staffs. No first-class flights or overpaid consultants. As I've mentioned before, living and working with integrity makes life a lot easier because you never have to answer for bad behavior. It can also lead to a lot of loyal support when you need it.

Finally, a concept for The New Renaissance Center began to coalesce. I would use my judgment and experience as a generalist to identify creative people with visionary ideas and fund their projects. I would emphasize environmental solutions but stay open to other areas as well. Before starting my search for these outstanding individuals, I gathered my funders together to explain my exciting plan.

They weren't excited at all.

One of them asked me an excellent question: "We know *you*. We've always liked *your* ideas. Why would we pay for other people's projects?"

The question stopped me cold. I realized at that moment that I should trust my own vision for a better environmental future, and start building an organization of my own. Why hadn't I thought of that in the first place? Was it a lack of confidence? Did my job in Sacramento get me stuck in the habit of delegating? I really don't know the answer, but it's a testament to the wisdom of my supporters that they steered me in the right direction.

The people at that meeting, all accomplished and committed environmentalists in their own right, formed the backbone of the board of directors of what would soon be known as the Resource Renewal Institute. Happily, most of them stayed with me through the years, providing the kind of perceptive advice that turned me around at that first decisive meeting.

Among them was Alf Heller. Brilliant, engaged, and farsighted, Alf passed away in the last days of 2019, leaving behind an astonishing environmental legacy as the founder of California Tomorrow, an innovative organization that led the way in managing the effects of California's explosive growth through comprehensive planning and conservation. In the late 1950s, he and Samuel E. Wood coauthored a sweeping report called *California Going, Going: Our State's Struggle to Remain Beautiful and Productive*, which proposed a statewide solution that challenged the establishment to implement his plan or come up with something better. A writer and editor above all, Alf was responsible for many books and periodicals about conservation. His wide-ranging interests included a lifelong passion for world's fairs and international expositions, subjects he wrote about extensively. Shortly before his death, he attended his twentieth exposition, this one in Milan, Italy.

Sitting next to Alf was Marion French Rockefeller Weber. A gracious person and insightful, invaluable board member, Marion is a longtime friend who has shown her support for my work in countless ways. Among her many philanthropies is the Flow Fund Circle, whose mission is to increase generosity, trust, discernment, and community. Marion continues a family tradition of committed environmentalism begun by her father, Laurance, who was so instrumental in helping me save Hawaii's Seven Sacred Pools back in the sixties.

Joining Alf and Marion was Dorothy Lyddon. Dorothy was a dear friend, lifelong feminist, and generous contributor who believed not only in RRI's goals and programs but also in my unorthodox way of doing things. I benefited greatly from her incisive and forthright assessments of our work.

In the years that followed, the original members were joined by other remarkable individuals:

Bill Bryan. In Chapter Seven, I described meeting Bill when he chose me as one of the subjects of his doctoral dissertation. Bill has a lifetime of experience founding and directing environmental nonprofits that emphasize the power of nature to elevate our human nature.

Yvon Chouinard. A great fishing friend and founder of the Patagonia outdoor gear company, Yvon brings to our board his knowledge and perspective as the foremost innovator of pro-environmental business practices. His corporate mission statement says it all: "We're in business to save our home planet."

Chris Erdman. Chris's extensive background in agriculture helps us understand the farmer's point of view and how it intersects with wilderness preservation.

Annette Gellert. A warm and upbeat person, Annette is a passionate environmentalist who directs a family foundation. Along with Peggy Lauer and Leslie Leslie, she created the WELLNetwork, an influential policy organization of women advocating crucial changes in California's toxic chemical laws and regulations.

Sylvia McLaughlin. Sylvia was one of the famous "three Berkeley housewives" who founded Save the Bay in 1961. Their homegrown movement took on the Bay Area's most powerful corporate, political, and real estate forces, eventually putting an end to all waste dumping and landfill development in San Francisco Bay. Since then, Save the Bay has been working to recapture and revitalize wetland habitats and transform garbage dumps into urban parks. Sylvia was an activist well into her nineties, famously sitting in a tree at age ninety-one to prevent the destruction of a Berkeley oak grove. She passed away in 2016 at ninety-nine. As a board member, she was a role model of persistence, inspiring me and everyone at RRI to keep going and never give up.

Sylvia McLaughlin, Resource Renewal Institute board member and cofounder of the transformational Save the Bay movement.

Ron Lovitt. Ron and I became close friends on the Larkspur Ferry, when we were both commuting to and from San Francisco every day. He was a successful attorney who needed an outlet for his fervent idealism. I was happy to jump in with some ideas. Over the years, his legal skills have served us in countless ways. One stands out: he played a pivotal role in a landmark class action suit – often called the grandfather of class action – that compensated consumers for the fraudulent real estate sales practices of the Boise Cascade company. It set a precedent that enabled many subsequent victories for environmental protection.

Bern Shanks. I met Bern when he was a ranger in Yellowstone National Park. We shared a rural Midwestern background and hit it off right away. Even as a young man, he so impressed me with his integrity and commitment to our national parks that, years later, I asked him to join my staff in Sacramento. I sent him to Washington, D.C., to keep a close eye on James Watt, Reagan's anti-environmental secretary of the interior. We called him our "Watt watcher." Later in life, he became director of fish and wildlife for the state of Washington where he came under tremendous pressure from timber interests to jeopardize salmon habitat. As always, Bern refused to compromise his principles. In an interview, he explained: "If I get tossed out on my ear because of that I'm willing to risk it. I reached a point in my life where hopefully this is going to be my last job and I want to do it with integrity, honesty, and openness." He was forced to resign, then went on to become a nationally recognized expert on dams, often engaging in projects with me and my team. Bern's last position before retiring was head of research at the US Geological Survey agency in Seattle.

Jocelyn Alexander Sladen. A resident of Alexandria, Virginia, Jocelyn traveled to California year after year for our board meetings. She shared her passion for wildlife conservation with her husband, William, a prominent British zoologist who was an authority on wild swans and birds of Antarctica, where a mountain is named for him.

Richard Silberman. I met Dick when we both worked for Governor Brown in state government. Our board benefited greatly from his political skills and financial experience.

There are other supporters who, while not board members, have been by my side throughout the years. I'm especially grateful to Gil Ordway and Helen "Babby" Dreyfus. Gil and I met on a Wilderness Society rafting trip up the Salmon River. He is one of the nation's most dedicated environmental philanthropists, serving on the boards of many prestigious organizations and endowing several scholarships at the Yale School of Forestry and Environmental Studies. That he would include my small nonprofit for support means the world to me. Babby Dreyfus buoyed us with her spirit and enthusiasm. She had a special interest in wolves and their reintroduction to the forest ecosystem and taught us a great deal about their intelligence and behavior.

Good friend and Resource Renewal Institute board member, Marion French Rockefeller Weber. She supported my work from day one.

Along with their individual talents and contributions, my board and other contributors have given me the two greatest gifts imaginable – trust and independence. That's how I've been able to pursue ideas I believe have great importance and wide applicability, but whose significance isn't always obvious at first. With my loyal board, I've always had free rein to do what I think is right.

So I'd say if you're starting a nonprofit, be sure to bring on board members you've known and trusted for some time, people who will let you do what you set out to do, who share your mission and want to serve it. In return, you must act responsibly, spend modestly, and stay financially scrupulous and transparent. One more piece of advice: be on the alert for people who want to use their board memberships for status and social climbing. They are not helpful.

Soon after my friends and future board members gave me carte blanche to form my own organization, I started experimenting with some projects.

One was Export Excess. It was conceived as a way to send used American farm and light industrial equipment to communities in developing Asian countries so they could become more self-sufficient. Looking back, it was a pretty good idea, not all that different from the microlending movement that was just starting around that time. Plus it had an environmental component – recycling.

I got in touch with Michaela Walsh, a delightful powerhouse of a person I had met when she was with the Rockefeller Foundation. Like our mutual friend and Nobel Peace Prize–winner Wangari Maathai, Michaela is a woman of firsts. The first woman to work in management at the Rockefeller Foundation. The first woman manager and then president of Merrill Lynch International. The first woman partner at Boettcher & Company investment banking firm. The founding president of Women's World Banking, the first and hugely successful global women's microfinance organization run by women for women. Today, Michaela speaks and consults all over the world. I love her insight into how to win over the most entrenched forces in society: "People who are not risk-takers will always challenge you in a way that if you challenge them back, they back away."

Michaela referred me to enterprising women in her organization who agreed to get involved in Export Excess. On my end, I made arrangements to fly to Bangladesh and meet with a potential associate who ran a number of sewing factories there. After my arrival, I checked into the Hilton and made my way over to one of his factories to meet him. As I was ushered into his office, I noticed two men with Uzis draped across their chests. It turned out I had shown up during a tense and violent labor dispute. The meeting went all right, but walking back to the Hilton, I heard gunshots – lots of them. I decided to take the next plane out of the country, regardless of its destination. When I called the airport to book a flight, I found out it had been closed down. Luckily, a kind man who worked at the hotel said he would notify me as soon as the airport reopened. After a while, he called my room and let me know that a few planes were taking off, and he was able to get me on a flight to Bangkok. I don't know what I would have done without his help.

After arriving in Bangkok and no longer fearing for my life, I started thinking of some way to salvage the trip to Asia. I remembered that my beloved 49ers were about to play in the Super Bowl. I called our embassy in town and talked to the guys in the Marine Guard, figuring they were likely to be football fans. I was right. They told me the championship game was going to be televised live at the Singapore Intercontinental Hotel. It would mean a two-and-a-half-hour flight from Bangkok. I thought, "What the hell," flew to Singapore, and took a cab to the hotel. By then, it was way past midnight and the streets were deserted. I walked into the empty hotel lobby and began to wonder if I had flown to Singapore for nothing. Then I heard cheering coming from the direction of the grand ballroom. The Super Bowl was in progress. Putting the last few days behind me, I joined the boisterous crowd and watched Bill Walsh and the 49ers defeat the Dolphins 38 to 16.

That trip killed any interest I had in Export Excess, although my partners continued with the venture for a while, backpacking around China and looking for people who could benefit from our equipment. For my part, it was a long way to go to watch the Super Bowl.

Back at Fort Cronkhite, other ideas came and went, some better than others. One I'm especially proud of was born over a campfire in 1981 along the Colorado River in the Grand Canyon. What brought our group of strangers together was the recent election of President Reagan who had famously said, "If you've seen one redwood, you've seen 'em all." For his new secretary of the interior, he had recently appointed James Watt, a Wyoming Republican with an even worse environmental record than his own. Our plan was to use our evenings strategizing ways to protect nature from the new administration's inevitable onslaught. Everyone in the group was asked to lead one evening's discussion. On the day of my talk, I thought about what topic I would address that night. I came up with something and presented it over dinner: "Let's form an organization that will protect the wildness of this magical place forever. Let's call it the Grand Canyon Trust." Everyone loved the idea and went home ready to make it a reality. Soon, GCT became its own nonprofit, not part of my organization, but I joined the fundraising efforts.

Susan Ives helped me mobilize donor interest in the Grand Canyon Trust, as she has done with so many causes during our almost four decades as close friends and collaborators. An environmental communications specialist, Susan is a master at framing, explaining, and publicizing critical issues to galvanize public support. When we first got to know each other, I could see the depth of her talent and potential and urged her to apply to the Harvard Kennedy School. In typical Susan fashion, she turned a class project into an ingenious solution for financing the restoration of the polluted, garbage-ridden Boston Harbor. Offered a position by the Massachusetts secretary of the environment to execute her plan, she was instrumental in establishing the public-private Massachusetts Environmental Trust and the vibrant harbor area that exists today.

With Susan's help and the work of many others, the Grand Canyon Trust sparked the imagination of many philanthropists. In just four years, it was officially launched at a festive party at the Museum of Natural History in New York City. NBC anchor Tom Brokaw acted as emcee and Arizona governor Bruce Babbitt was the keynote speaker. Today, the GCT is going strong, having expanded its purview to include not only the canyon but many other national parks and natural monuments in the Southwest.

Soon thereafter, I was hired by Will Hearst, the grandson of newspaper publisher William Randolph Hearst, to write a bimonthly opinion column on the environment for his daily newspaper, the *San Francisco Examiner*. I knew I would never lack for subjects to rant about, but I decided I needed an editor to whip my writing into shape. Don Terner, a great guy and visionary environmental architect, planner, and Jerry Brown's housing director while I was secretary of resources, recommended Peggy Lauer, a former journalist who worked at *Mother Jones*. She wanted to focus her work in the environmental movement.

Something I never anticipated when I took on the assignment was how much *I* would learn from nearly a decade of writing the column. It's one thing to have a strong opinion about a subject. It's quite another to organize, support, and articulate it convincingly. It usually took me about three thousand words to do that. Unfortunately, the *Examiner* has a six-hundred-word limit. That's when Peggy would come to the rescue. Like most writers, I hated giving up even one precious word. To get me to go along with her well-considered edits, Peggy had to be far more persuasive than any of the columns we worked on. Over time, I began to understand that the often-painful process of writing and editing was actually turning me into a clearer thinker. The lessons I learned as a columnist have served me well in crafting the many speeches, articles, and books I've authored in the years that followed.

Looking over the columns now is like thumbing through a scrapbook of what matters to me most. Saving irreplaceable open spaces from development. Protecting wildlife and their habitat from degradation and destruction. Preventing private interests from plundering our public lands. Keeping natural resources sustainable for generations to come.

Rereading what I wrote thirty some years ago also reminds me of how things get accomplished. Some all at once, others over years and even generations. Still others move two steps forward and one step back. Through it all, it's critical to keep the public informed because, by and large, most people want to be environmentalists.

It didn't take long for Peggy to become much more than an editor, although she did that exceedingly well. In fact, she became my indispensable collaborator and close friend. For most of her twelve years at RRI, Peggy's title was vice president of programs, but I would describe her as vice president of everything. Peggy used to say that her job was putting legs on my ideas, a far too modest description. With her creativity, clarity of thought, quick wit, and work ethic, Peggy brought those ideas to life, explored their possibilities, and realized their full potential. She also contributed many more of her own, helping to shape RRI's direction for years to come. But what I enjoyed most about Peggy, and still do, was her energy and buoyant spirit. I think it has something to do with growing up with a sister and *six* brothers. Nothing got the best of her. She could handle whatever came her way.

As the days and weeks passed in my little office on the beach, something began take shape that felt true, that had weight. My thoughts returned to Investing for Prosperity, the unified natural resources policy plan I had developed for the state of California. By integrating the isolated and often rival departments under my authority, I had saved the taxpayers over a billion dollars in the five years of my tenure, got industry onboard to replenish the resources they profited from, and transformed California from an exploitative to a sustainable resources manager. I also seemed to have a real knack for getting contentious factions to somehow agree on big issues.

At the same time, I remembered the international conferences I had attended as secretary, where I discovered that a few other countries were thinking along similar lines. They were also attempting to move beyond old patterns of conflict and litigation to a proactive collaboration among business, government, the academy, and environmentalists.

I remembered a term I'd heard at one of those conferences. It captured this elegant, intelligent, flexible, and forward-thinking approach in one word – "Greenplanning."

I wondered: what if I took the lead in spreading this potent concept to governments around the world?

I was fired up and set out to start a global movement to bring Greenplanning to every country in the world. A path forward began to emerge. First, I would make fact-finding visits to countries where Greenplanning programs were already in use. I had heard a few things about Canada and Norway. Maybe there were others. Second, I would integrate what I learned abroad with my own experiences in Sacramento and put together educational materials and a blueprint for Green Plans. Next, I would use my sales know-how (it always comes in handy) to bring more nations on board – holding international conferences, meeting with governmental environment and resources leaders, talking to the press. It was at this point that I changed the name of the organization from New Renaissance Center to Resource Renewal Institute. to align with my new Greenplanning direction.

Back in the days before Google, Peggy and I started searching high and low for any conferences or meetings we could attend relating to comprehensive environmental strategies. There wasn't much to be found. Then Peggy happened to see a small article in the back of a newspaper. It said that the Canadian government was hosting a conference and inviting environmentalists from other nations to attend. We were thrilled. At last, we could meet with like-minded people and gather information to move our work forward. We signed up immediately.

When we arrived in Ottawa, it was freezing cold. With a little time to kill, we borrowed ice skates and headed down to the frozen

Ottawa River that flows through town. We joined commuters skating home from work and had a lovely time. Later, over dinner, we diligently prepared for our big meeting the next day. Arriving at a stately government building in the morning, we entered a palatial hall, eager to jump into a day of talks with our fellow environmentalists. We were warmly greeted by a delightful official who ushered us into a meeting room where five or six people were seated around a small table. We were confused. Where was the bustling throng of people representing nonprofits from around the world? We soon found out that the meeting had been organized for just a few top-level environmental officials from Canada, Norway, and the Netherlands, the leading nations in forward-thinking environmental policy. But because we had expressed such enthusiasm during our earlier phone call to the Canadian hosts, they kindly invited us to join. It was a remarkable happy accident that led to our forming close relationships with the world's leading environmental ministries.

Lesson Learned: You never know where something will lead.

Soon thereafter, Peggy went to Norway, where she met with environmental leaders and learned a great deal more than we had in Canada. The best advice she received in Norway was to go to the Netherlands. Within a few weeks, I was on my way there. The Norwegians were right. The Netherlands was far and away the most advanced Greenplanner in the world. I went back a number of times and gathered a huge amount of information that was key to our building a better model. In fact, RRI organized a number of Seeing Is Believing tours to the Netherlands to educate environmentalists and government officials from the United States and around the world.

There was a good reason why the Netherlands had taken the lead in Greenplanning. With its small size, dense population, and heavy industry, the nation had become the most polluted in Europe. Their backs against the wall, the Dutch had no choice but to work together toward a solution. In doing so, they set an example for the entire world. The Dutch Green Plan – agreed to by all stakeholders – set extremely high environmental standards for industry, but let each business figure out for itself how to reach those goals. This approach eliminated onerous regulations, freeing up companies to innovate solutions on their own. Before long, Dutch companies had gained a competitive economic advantage in environmental recovery methods over those in other countries. In addition, companies and consumers were held financially responsible for the full life cycle of the products they made and bought, including recycling. For example, the cost for the responsible disposal of a car was built into its initial price.

Peggy and I also traveled to New Zealand, where recycling and composting had long been part of daily life, and bitter disputes over old-growth logging were finally being resolved through tense yet successful negotiations among rivals. The New Zealand government decided to scrap hundreds of redundant and wasteful governmental agencies and divisions, replacing them with fourteen streamlined departments organized by watersheds. Like the Dutch, New Zealand businesses became environmental innovators, earning money by exporting their cutting-edge technologies – yet another example of how smart environmental policies can be economic drivers.

In Singapore, I saw how a crowded urban city-state on a small island was creating and connecting beautiful parks and open space while requiring all new construction to be green. These programs continue today, with additional mandates that all new buildings incorporate elements like green walls, roofs, and vertical gardens. Updated in 2012, the Singaporean Green Plan ushered in even higher standards of clean water and air and energy efficiency.

Not everything in these countries was or is perfect, of course. The important thing is that, decades later, these national Green Plans have proven to be workable blueprints that adapt to changing times and challenges. That's saying a lot.

Armed with these examples of successful Green Plans, we came back to the States, did extensive writing, and felt ready to hold the conferences we knew were key to initiating and supporting programs around the world. Now it was time to raise the money we needed to make it happen. We heard tell that the famously wealthy Buck Foundation was about to donate millions to three yet-to-be-chosen Marin County nonprofits through the newly established Marin Community Foundation. The deadline, however, was just a few days away. We decided to go for it. Pandemonium ensued, but somehow Peggy learned how to use a MacIntosh for the first time, and we put the damn application together. On the day before the document was due, she stayed up all night, then drove like a madwoman, arriving at the foundation headquarters just under the wire. We were rewarded with a happy ending – $500,000 to mount three global Greenplanning conferences at the Marin County Civic Center Auditorium.

As always, there was no time to wallow in our success. We had to begin planning the content and logistics of the meetings, which were only about four months away. Peggy and I expanded our team to include several bright and dedicated people like Joe Seton, Mark Valentine, Jennifer Carroll, and Darcy Rollins. Darcy had been a star soccer player, and it showed in her focus and tenacity.

Even more challenging than planning the event was getting people from around the world to attend. Most had never heard of Resource Renewal Institute or Green Plans, so there were few signups at first. Then a young woman named Susie O'Keeffe walked into my office expressing a desire to help the environment. I immediately put her on the phone to get us some participants. Susie was a real go-getter and a natural salesperson.

She signed up most of the attendees, three hundred government officials from the United States and countries in Asia, Africa, and Europe. Like so many of the others who joined RRI when we were preparing for the conferences, Susie stayed on and became an important part of our operation. She eventually moved to France, where she coordinated our Seeing Is Believing policy tours in the Netherlands.

At the conferences, our colleagues from the Netherlands and New Zealand joined us in leading the sessions, explaining how their national Green Plans worked and could be adapted to different climates and types of government. Two of the most magnetic and valuable speakers were Tom Fookes from New Zealand and Hans Van Zijst of the Netherlands. Tom was both a specialist and a generalist, studying the intersection of human habitation and the natural world. He authored much of New Zealand's Resources Management Act, one of the most influential in the world, and innovated a public participation program that involved ordinary New Zealanders in the process of redesigning the nation's environmental policy. Tom continued to partner with RRI on Greenplanning for the rest of his life, informing our work in countless ways. Hans Van Zijst played a similar role in the Dutch government, finding creative ways for business and environmentalists to cooperate. He and his colleagues Herman Sips and Paul deJongh were instrumental in the Dutch National Environmental Policy Plans 1, 2, 3, and 4, which helped lay the foundation for Green Plans around the world.

As we had hoped, these initial conferences proved to be springboards for additional international meetings in Sweden and a number of other nations. Over the next few years, we developed an archive of materials to assist governments, businesses, nonprofits, and communities in building their own national Green Plans. We produced a documentary and published regular newsletters incorporating the latest knowledge from programs around the world. Vice President Al Gore became an advocate. The International Network of Green Planners was formed. It was clear that Greenplanning was growing into the global movement I had envisioned. Our tiny nonprofit had become a trusted advisor to many national governments, helping them create or expand integrated natural resources programs that would sustain them for generations to come.

One of our most exciting collaborations was with the national government of Mexico and its capital, Mexico City. Their leaders were enthusiastic supporters of creating a Green Plan that would provide both environmental protection and economic growth for the Mexican people. Danielle Kraaijvanger, who had originally joined RRI to support Peggy's work, was fluent in Spanish and played a key role in shaping the Mexican Green Plan. An organizational wizard, Danielle later succeeded Peggy as RRI's vice president.

Before long a comprehensive plan was underway, one that would serve as a template for Greenplanning in developing nations. At the same time, I focused on getting US state governments involved in Green Plans. Our first success was the state of Minnesota, which brought in 3M and other large corporations headquartered in Minneapolis.

At this point, I'd like to explain more about Greenplanning, and why, all these decades later, I still believe it holds the key to solving the existential challenge our world faces today – climate change. Greenplanning follows the simple principle that to solve anything you have to solve everything. Not only that, solving everything is actually much easier to achieve than solving one thing.

As a confirmed generalist, I have seen time and again the disastrous consequences of specialization and the blinders it imposes on even the most brilliant minds. Greenplanning, on the other hand, is simple and all-inclusive. It doesn't require any scientific breakthroughs or technical innovations, just the willingness of all parties to participate in a comprehensive, far-reaching solution based on natural systems. And since it is executed by governments with the participation of all of society's stakeholders, a Green Plan doesn't even depend on human beings acting any more virtuously than we do now. On the contrary, industry leaders, environmentalists, and other participants are expected to fight tooth and nail for their own interests during negotiations. They just have to abide by the final contract.

Especially in the face of climate change, it is clear that the world should have an Earth Green Plan. But until we humans are willing to embrace that reality, another genius of Greenplanning is that it can be modular and grow incrementally – for towns, cities, states, regions, nations, or continents – and then connect over time.

I wrote a whole book about the subject a couple of decades ago titled *Green Plans: Greenprint for Sustainability*, with a beautiful foreword written by my friend and hero, David Brower. It has been updated with two more editions and is now a well-known course book for college classes, a guidebook for governmental policy makers, and a useful tool for corporate environmental executives. I'm proud to say it is written in an accessible style, making it a worthwhile read for anyone interested in the earth's future, as we all should be. If you'd like to learn more about how Green Plans work, I recommend it.

What's happened to the Green Plan programs we fostered and supported in the nineties? Many, like those in the Netherlands, New Zealand, and Singapore, have become an established and accepted part of their government infrastructure. In fact, some of the Dutch Green Plan policies have been adopted by the twenty-seven nations of the European Union. Other plans have been the victims of shifting political winds. Vicente Fox, for example, was elected president of Mexico in 2000 shortly after we completed our Mexican Green Plan, and he gave in to pressures from special interests to kill it. Likewise, in a surprise victory, the wrestler Jesse Ventura became governor of Minnesota in 1999 and put an end to that state's developing Green Plan. For now, Canada is a mixed bag. The national program was suspended when a conservative government gained power, but a number of provinces have maintained their Green Plan programs.

That's the way it is with environmentalism. Politics intervenes. Unforeseen events happen. You have to be patient and resilient, and always persevere. Eventually you win. I'm living proof of that. Peggy has pointed out to me that whenever we encountered a major setback in our work at RRI, I would always say the same thing: let's get on the phone. It was and still is my way of immediately changing a negative reality. I would get on the phone to raise funds, get advice, contact an influential ally, inform the press, write an op-ed, lodge a complaint, file a lawsuit – whatever it took to regain some control. We weren't always successful, of course, but more often than not, we were. And if our actions didn't fix the particular problem at hand, they invariably led us in an important new direction.

Another way I like to look at obstacles is to picture a dog with a bone. She'll happily gnaw and chew on it for a time, then after a while, she's no longer getting what she wants from that bone. So she turns it over, flips it around, stares at it from different angles, sniffs it here and there. It's the same for human beings. If we look at our problems from different angles, rethink and reframe what we're dealing with, we usually find an answer.

The Greenplanning concept has proven its worth time and again, and it will prevail. Just look at the Green New Deal that came to light in 2018. It may have a slightly different name and some new angles, but it is essentially another version of Greenplanning – an integrated, comprehensive, cooperative, government-driven system to tackle climate change and protect natural resources.

In the big picture, this is how I see it. Old ideas and new ideas are always competing with each other. At first, old ideas seem to have all the power. They are safely in place, comfortable, familiar, with lots of friends hovering around to support them. New ideas are foreign and hard to grasp. They seem lonely, with few allies to speak up for them. But when we humans finally realize that we desperately need what the new idea offers us, we embrace it surprisingly quickly and easily. Think of democracy. Philanthropy. The scientific method. At one time, these were alien, far-fetched concepts. Today, we can't imagine life without them.

In 1996, the Clinton administration recognized RRI's achievements in Greenplanning as well as its beginnings with Investing for Prosperity, the resource sustainability infrastructure I established for the state of California. Along with the California State Resources Agency, we were presented the President's Sustainable Development Award by Vice President Al Gore, our most environmentally minded leader before or since. The award demonstrated a real understanding of sustainability by calling out its positive effects on job growth, economic productivity, and recreational opportunities.

Along with Greenplanning, I continued exploring other themes at my environmental incubator, Resource Renewal Institute. My only criterion was that every new idea have widespread value and application leading to a healthier future for the planet. It could be a roadmap for better governmental resource policy, or a replicable pilot program, or an educational model for nurturing tomorrow's environmental leaders. There were no restrictions. It just had to work.

In the late eighties, for instance, I was looking for a project that would offer something useful to the delegates attending the 1992 United

Nations Earth Summit in Rio de Janeiro. I landed on creating a local version of Investing for Prosperity. I thought it could benefit Marin County and also serve as a template the Earth Summit participants could use in their own countries.

Heading the project was Andy Russell, a young man who later went on to a career at the Corporation for Public Broadcasting and PBS affiliates. First, Andy researched how Investing for Prosperity had been doing in the years since I left Sacramento. The answer was very well, so we felt confident recommending its methods to others. With its mix of national, state, and local parks; mountains, rivers, and coastline; significant agriculture; transportation systems; and urban, suburban, and rural communities, Marin proved to be a good microcosm for the project's larger purpose. And just as we had discovered in Sacramento, the lack of integrated organization and resource management in Marin had led to the same problems we encountered statewide – waste of money and natural resources, redundancy, and departments working at cross-purposes. We graded various functions, departments, and regions in the county from A to F. Point Reyes received D minus for their neglect of the park's land and wildlife, and we became the subject of a scathing editorial in the local weekly, the *Point Reyes Light.*

Although our Marin County project had gone well, I lost all interest in attending the Earth Summit once I found out that President George H. W. Bush refused to attend. Frankly, both Peggy and I were embarrassed to represent a nation unwilling to participate in such a significant gathering. I called my friend Wangari Maathai, the courageous founder of the Kenyan Green Belt movement, and told her we wouldn't be seeing her in Rio after all. I describe Wangari's creativity and tenacity in Chapter Six, and she demonstrated those qualities that day on the phone. "Why don't you join our Kenyan delegation? I'll sign you up." Then the *San Francisco Examiner* handed us press passes, which gave us carte blanche to attend any meeting we wanted to.

I'm glad Wangari convinced me to go. The Rio Earth Summit was well worth attending. The delegates took their work seriously, resulting in several global agreements and the founding of the Green Cross, an international organization led by Mikhail Gorbachev, charged with implementing the protocols established at the conference.

As time passed, Resource Renewal Institute proved to be a nourishing environment for effective, far-reaching ideas. Personally, I flourished in its intimate, casual atmosphere, supported by a few like-minded generalists who were willing to go beyond the usual nonprofit mold, to think big and do whatever it took to make things happen.

The People Who Made It Happen

I've been told I have quite a knack for finding great folks to work for me, some accomplished veterans, and many who had more potential than experience. I don't know whether I have that ability or not, but I do know that the people who have contributed to RRI deserve most of the credit. They are an exceptional group – inventive, multitalented, committed, and kind.

Following Peggy and Danielle were vice presidents Allison Jordan, then Elizabeth Baker. Allison picked up the Green Plan baton, developing a universal Green Plan manual, organizing dialogues between California and the Netherlands governments, setting up trips to New Zealand and summits in the Dominican Republic, partnering with Jerry Brown when he was mayor of Oakland to integrate the concept into his city planning, and exhorting every state in the country to adopt California's more stringent auto emissions standards. The list goes on. She accomplished all that in her seven years with us, then moved to the Wine Institute as head of environmental affairs.

I was especially gratified when Allison told me that she always felt I supported her as a mother of small children, encouraging her to take the time she needed for a school conference or a sudden illness. She asked me once why I was so sympathetic to her situation, unlike other bosses she had had. I had to ponder her question for a while because, frankly, I had never given my behavior a moment's thought. It just came naturally to me, and having been a father of young kids myself, I knew the territory. The conversation went on, with Allison wondering why I had always hired women in leadership roles and given them opportunities they had been denied elsewhere. As an example, she mentioned the women I had chosen for my internship program in Sacramento. Again, it wasn't anything I consciously set out to do. I have always looked at human beings as individuals, and if I encounter one – man or woman – who deserves a chance, I find a way to open a path forward. After all, I'm the one who ultimately benefits from their success.

On the lighter side, I remember one morning when Allison arrived at our office on the San Francisco waterfront. I was meeting with someone at the end of the pier and motioned to Allison to walk over and join us. As she got closer, it dawned on her that the person I was talking to was Robert Redford, an avid supporter of environmental causes. She was flabbergasted and unable to utter a word at first, but she regained her composure and picked up on the conservation quite well.

Elizabeth Baker joined us when Allison left. As an undergraduate at Williams College, she had initiated a project that transformed the university's meager environmental archives into a comprehensive collection. She was full of new ideas for us too, and I invited her to join our team. She brought a unique intelligence and creativity to our efforts, exactly what is needed at an organization whose stock in trade is global strategic thinking. A true Renaissance woman, Elizabeth left us after a few years to pursue her first love – opera. Today, she sings with the San Francisco Opera. Elizabeth once kindly said that anyone who works with me should pay tuition. I could say the same about her.

When I look back at all those who have worked, consulted, and volunteered at RRI over three decades, I see a cast of fascinating characters, each one contributing something different and distinctive to our mission and our lives. I'm pleased I've been able to enable some younger people to appreciate their own potential and pursue careers that have served them and the environment well. There have also been many at the end of their careers who donated experience, expertise, and wisdom we could never have afforded to pay for. I often refer to our group as my cadre, a circle of devoted and knowledgeable friends who come and go, expand and contract to do what needs to be done. A word about some of our staff I may not have mentioned before:

Jean Wetzel Chinn started out in a clerical role at RRI. She discovered a passion for environmentalism, decided to study biology at UC Berkeley, and became a biologist with the California Department of Fish and Wildlife.

Marilyn Price was our office manager and IT person in the days when we had one computer with a tiny green screen. She rode her bike to work rain or shine, eventually combining her passion for mountain biking with a desire to help underserved kids by starting her own nonprofit, Trips for Kids. What began as one bike trip for children without access to bikes or open space is now a thriving organization that has served 140,000 young people. Trips for Kids also manages a bike shop where young volunteers learn how to repair bikes, earning points to buy their own bikes and parts.

Jason Kibbey already had a degree in environmental economics and policy from Berkeley when he joined our team to begin our Defense of Place program. He is now CEO of the Sustainable Apparel Coalition, a worthwhile nonprofit working with clothing manufacturers to reduce their impact on the planet. Jason is married to Elizabeth Baker.

Lynn Alexander brought valuable research skills from her time at FEMA. She investigated and uncovered the vast amounts of money owed to the American people by private businesses that use public resources and never pay for them. She later brought the same rigorous standards, determination, and thoroughness to our Forces of Nature archives.

Forces of Nature also benefited from the taste and sensitivity of video editor Lauren Veen and videographer/editor Vincent Tremblay. In vignettes of less than five minutes each, they showcased the wisdom and vision of more than 160 environmental pioneers, documenting each remarkable story of courage and perseverance.

Laci Videmski played an indispensable role in developing RRI's Water Atlas, using his sophisticated technology expertise to reveal and clarify how California water is used – and misused – by special interests. Laci also did a great job getting RRI's Forces of Nature website up and running.

My son, Tyler Johnson, developed RRI's own website. Back then, few nonprofits had websites, and Tyler built an impressive one that advanced our reputation as an effective and forward-thinking organization. He had graduated from UC Santa Cruz with a degree in environmental studies, so he also brought a wealth of knowledge to the programs we were in the midst of developing in those early days at RRI, especially Greenplanning.

John Levinsohn is a philanthropist who motivated us all with his energy and enthusiasm. He did much of our legwork on European Green Plans, helping us lay the foundation for our own Greenplanning program.

I was also fortunate to find financial advisors as concerned about our cause as they were about our money. John Skov, one of my best friends for many years, was a welcome presence when he appeared at the office several times a month. When he passed away, John Angel followed in his footsteps, contributing his lively mind and cheerful personality along with his meticulous care of our books. Like his predecessor, John shows great respect for the fact that the money in our budget derives from generous individuals and foundations we must always be accountable to.

There were and are many more invaluable contributors. Among them Corey Kirkwood, who was a great help to Danielle and me in putting together the Mexico Green Plan; Leslie Leslie, our exuberant volunteer who recruited hardworking interns; and Ted Dunham, who later became an executive at GE in Seattle, also contributed significantly during his time at RRI.

Today, RRI is smaller than in the past. I currently work with a staff of only two, but their commitment and versatility fill the shoes of many more. Our first president, Deborah Moskowitz, joined RRI after a career in public health. The same qualities that led her to pursue that field have make her ideally suited for RRI – a scientific mind, the tenacity to find the truth, a collaborative spirit, and a deep concern for the welfare of others. Although Deb modestly maintains that she sees her job as carrying my spear and enabling me to do my best work, she has brought her own exciting ideas and perspectives to our programs, improved them, and taken them in new directions.

Deb tells me that RRI enabled her to combine her two great passions – science and the outdoors. She grew up in Danville, California, when it was still more rural than suburban. As a young girl, she wandered the open fields, hiked and biked the trails of Mount Diablo, and dreamed of becoming a forest ranger when she grew up. In high school, she loved studying natural sciences, especially biology and physiology, and took a special interest in fish. This led to degrees in biology and public health from UC Santa Barbara.

I was introduced to Deb by her close friend, Nancy Graalman. A transplant from rural Oklahoma, Nancy is one of RRI's most dedicated volunteers. When I first met her, I was fired up about some shenanigans being perpetrated by the board of directors at the California Academy of

Sciences. A family had donated a lovely piece of land in Sonoma County called Pepperwood on condition it remain wild. But when the academy was low on cash, the CAS board decided to sell Pepperwood. In one of my columns for the *San Francisco Examiner*, I urged readers – including most of CAS's major donors – to refrain from giving money to the academy because they didn't live up to their agreements. No one had ever criticized this august institution before, so the column caused quite a stir. In the midst of it all, I got a call from a stranger who wanted to help me save Pepperwood. It was Nancy. She jumped in and worked like hell to protect the land. The academy's embarrassed board of directors finally backed down and went so far as to fire their director. Today Pepperwood Preserve is owned by a responsible foundation and remains 3,117 acres of wilderness.

Nancy had become a trusted colleague, so when she suggested I speak to her friend, Deborah Moskowitz, about an open position at RRI, I took her recommendation seriously. And thank goodness I did.

When I interviewed Deb, I was struck by how similar her aspirations were to ours, how much she wanted to take on big challenges and build a better world. What I didn't know at the time was how instrumental she would be in building a better Resource Renewal Institute.

Over the years, I always tried to run RRI as a place of openness – to all sorts of people, ideas, opinions, opportunities – as long we kept working toward the same environmental goals. I'm convinced we do that at a higher level since Deb joined us. Regardless of how busy she is, she finds a way to give every person and every idea time and consideration, to make suggestions that add value, and to lead with kindness and encouragement.

Like Deb, Chance Cutrano, RRI's director of special programs, is an avid generalist. A philosophy major in college, he was introduced to environmentalism by a philosophy professor who wisely viewed climate change as an existential issue. During a worldwide internship on innovative practices to mitigate climate change, he crossed paths with Michael Shanks, the son of my friend, board member, and colleague, Bern Shanks. In corresponding with Chance, Bern noticed an uncanny similarity between what Chance was doing as an intern and many of our programs at RRI. Bern put us in touch. It took a long while for us to connect, but Chance demonstrated surprising persistence, always an important quality to me. To prove his commitment to our cause, he camped out in our office and took the opportunity to read through many of the books in our library. A while later, after he had been hired, he showed me an email he had received from Bern. That's when I understood why he was so eager to join RRI. The note is one of the kindest things anyone has ever said about me, and I appreciate it no end. It also describes quite well how much I like mentoring promising young people and showing them what they are capable of.

> *I cannot fully express my admiration for Huey Johnson. We shared a panel debate in Butte, Montana over 35 years ago, both of us opposing the privatization of federal lands. He was then Secretary of Resources for California and hired me a few months later. He changed my life, my career and became a lifelong friend. I could provide you with dozens of examples of his legacy, including creating major national parks, protecting millions of acres of wilderness, founding the Trust for Public land where he acquired a million acres from pocket parks in urban ghettos to ecological gems. He earned the highest environmental award in the world but never looks back, never dwells on his accomplishments, only moving ahead toward his new vision. If we built monuments to environmentalists like presidents Huey would be comparable to John Muir, Teddy Roosevelt, Aldo Leopold.*
>
> *If this works out for you, let me know if I can help with anything. None of us who have worked for Huey would say it was always easy. But he has great sense of humor, is intensely loyal to*

those who work for him and a vision unlike any other environmental leader. If you spend some time with Huey Johnson you will be telling your grandchildren about the experience.

It has given me much pleasure to see Chance grow from a young intern to a multidimensional director of special programs. During his time with us, he earned a master's of public policy degree in sustainable management from the Presidio Graduate School. His work there benefited us as much as it did him, probably more. Fortunately, Chance, like Deb, brings a passion for research and getting to the bottom of things, something I've never had a flair for. And he is a tireless participant in local politics affecting the environment, attending all those meetings and hearings where decisions actually get made.

Both Deb and Chance work seamlessly with the RRI "cadre" I mentioned earlier, that circle of lawyers, specialists, friends, supporters, even graphic designers and artists, who seem to show up at precisely the moment we can use their help and expertise.

| *The dedicated staff of Resource Renewal Institute.*

RRI's Programs: What Our Think Tank Thought Up

I'm pleased to report that just about all our programs and projects have stood the test of time, often anticipating and influencing the most critical environmental concerns the world faces today. And while we talk about our programs as separate entities, they, like the environment itself, are interconnected, with one often leading to an idea for another. I am profoundly grateful to our funders who patiently go along with our organic way of working, trusting that the results will benefit the earth in important ways. I believe they do.

A number of our programs are related to water because California, in particular, has a woeful track record in this area. So if we can shake things up in our state, it can serve as a model for other states and countries, especially as climate change worsens worldwide drought conditions. The relationship of California and its water has filled many books, but the upshot is that water interests in agriculture and Southern California cities have always had a stranglehold on Sacramento policy. Public water is bought and sold by private farmers and ranchers as if they owned it. The use of groundwater is unmonitored and unaccountable. The ecosystems of countless streams, rivers, and estuaries have been subverted. And astonishingly, there is almost no established law protecting public water as there is for public lands. In fact, until very recently, "wild water" – the lifeblood of fish runs, riparian vitality, clean ocean water, migratory bird flyways, wetlands, and a profitable tourist industry – was considered without value, legally speaking, unless it was deployed for agriculture or the taps of LA and San Diego. Other states adjudicate water, requiring claimants to appear in court to certify their claims. Not California. The lack of oversight of groundwater, in particular, is an outrage. Right now, about a third of all drinking water in our state is groundwater. The extraction has left the Central Valley aquifer collapsed on itself so it can no longer absorb rainwater and recharge.

Shaken by drought and wildfires, the public has finally begun to take notice of these abuses, voting for statewide propositions that are shifting the power, at least to some degree. And today's Department of Fish and Wildlife seems to have brought in more enlightened leadership than in the past. I only hope the progress continues.

I like to think the inventive programs RRI has created to change California's water policies have helped to start turning the battleship around. We thought up The New California Water Atlas (NCWA) during one of our lunchtime brainstorming session as a way of shining a light on California's water records, which are purposely kept arcane and indecipherable by the powers that be. We went to work like a team of investigative reporters, teasing apart decades of facts and figures from dusty state records. The result is a free, online, searchable database offering instant access to information on California water. Now anyone can find out who is irrigating crops, providing stock water for animals, using water for mining, or sending water to cities. NCWA users can also determine where groundwater supplies are healthy and where they are threatened. If knowledge is power, the Atlas is the first and only tool that gives power to journalists, environmentalists, and the public in the battle for California's scarcest and most sought-after resource.

The Water Atlas project motivated me to resurrect an idea I'd had way back in my Nature Conservancy days: since I had been able to convince the IRS to give land contributors a well-deserved deduction on the property they donate, why couldn't the same principle apply to donated water rights? So around a decade ago, I decided to give it a serious try, calling it the Instream Water Transfers Project. RRI formally requested that the IRS issue a binding Revenue Ruling regarding the tax deductibility of permanent donations of appropriative water rights. We garnered the support of US senators, regional water and land trusts, western state water and natural resource agencies, and right-minded private landowners. Around the same time, we joined forces with American Rivers to form the California Water Trust Network to coordinate state and federal water policies and encourage voluntary water transfers. The CWTN has grown to include more than a dozen organizations who are now collaborating at the state level. At the same time, the coalition has found water-rights holders willing to be "test cases" by donating all or some of their water rights. If and when the judgments go the right way, these cases will establish a conservation tax precedent benefiting both the donors and the public, and leading to healthier streams, creeks, and rivers, and the fish, wildlife, and human beings that depend on them.

My friend Yvon Chouinard, who founded the Patagonia company, maintains that every good idea he's ever had came to him while he was fishing or hunting. The same is true for me, at least when it comes to Fish in the Fields, RRI's most ambitious program related to water, and probably our most far-reaching concept since Greenplanning

I was hanging out in a Central Valley duck blind during hunting season – relaxed, content, peacefully waiting for a flock to fly by. As part of a joint program of rice farmers and the California Department of Fish and Wildlife, many farmers along the Pacific Flyway flood their fields after harvest to promote the decomposition of rice straw and provide wetlands for migratory birds. As I sat waiting for the ducks, I began to wonder. What if all those fields of water could be put to another use, like shoring up the beleaguered California salmon population, made weak and scarce by dams and water exploitation? If the fish could linger in the flooded areas for a while, perhaps they would grow larger and more likely to survive their journey downstream. Before long, RRI had launched the Nigiri Project, named for the sushi dish made of rice and fish. The Nigiri Project proved quite successful, and eventually we transferred the project to the nonprofit California Trout (Caltrout) and UC Davis to do further research and implementation.

Still, I kept thinking about all those flooded rice fields and how they might somehow serve a larger purpose. As climate change and its impact were becoming more urgent, I felt the need to concentrate on a solution that could be scaled worldwide, like Greenplanning. Back in the duck blind, I let my thoughts wander beyond the Central Valley plains to the oceans, where uncontrolled trawler fishing and pollution wreak havoc on the ocean food web. Even when large fish are protected, the overfishing of the forage fish they feed on has the same destructive effect. It occurred to me that salmon might not be the only fish that could thrive in the nutritious soup of a flooded rice field. If a flourishing aquaculture of forage species could be developed in the world's rice fields, it could allow ocean forage fish to play the role that nature intended. The small fish grown in fields could be used for bait, aquaculture fish food, pet food, and maybe even human food.

(I'm embarrassed to admit that as I was patting myself on the back for my original concept, I had no idea that traditional cultures in Asia, Africa, and South America have been cocultivating rice and fish for thousands of years.)

The more I thought about it and talked with my staff and others, the more benefits the idea seemed to have, even beyond ocean preservation. Small fish could provide a new source of protein for a growing population. They could add a profitable and sustainable second annual crop for rice farmers, with no need for additional water, land, or food. And with another crop to harvest, farm workers and their families could stabilize in a single community rather than migrating from one location to the next.

I felt Fish in the Fields was ready to take off. We could start out in California's half-million acres of rice fields and adapt the model worldwide, where more than half the population eat rice every day. But then, like many new ideas, ours hit a big snag. One of the charitable foundations we had hoped would fund Fish in the Fields asked us how an environmental organization could justify focusing its efforts on the only plant-based agriculture that emits methane, a greenhouse gas twenty-five times more harmful than CO_2.

Good question.

Putting her scientific training and dogged curiosity to work, Deb Moskowitz began investigating the intersection of fresh water, rice, and fish. She scoured scientific journals for months and finally landed on a lead. A University of Montana aquabiologist, Dr. Shawn Devlin, had conducted research on methane in freshwater ponds in Finland. He found that the presence of fish interrupted methane emissions from ponds. Rice had no role in his research, but that didn't stop Deb from seeing an opportunity. She contacted Shawn, and before

Fish in the Fields experimental ponds, where we proved that introducing fish into fallow flooded ricefields reduces methane emissions by more than two-thirds.

long, RRI was spearheading a pilot project near Marysville, California, to see if adding fish to rice fields could significantly reduce the crop's methane emissions.

The research was rigorous and the results groundbreaking. Our little nonprofit – not a well-heeled university or corporation – proved that raising fish in fallow flooded ricefields reduced methane emission by two-thirds and probably more. Scaled across the world's four hundred million acres of rice farms, the Fish in the Fields concept could play a major role in mitigating climate change.

Fish in the Fields has been causing excitement and gaining interest among a wide range of people and organizations. And we've received exceptional guidance and advice on working with state agencies and officials from none other than Jonas Minton, who, as a committed young attorney in Sacramento in 1972, gave up his Christmas holidays to help me win passage of the RARE II federal wilderness legislation.

In the last several years, Fish in the Fields has grown from a one-acre experiment to a hundred-acre pilot project attracting interest from around the world. Exactly where it will go or what it will become isn't clear yet, but I know its success will mean more sustainable oceans and cleaner air.

I've described at length the many years I spent as a land saver, a profession my colleagues and I pretty much invented. But after a while, it became clear to me that saving land is only half the story. As time passes, a parcel worth $1,000 when it was donated might increase in value to a million or more. That creates a stong temptation for city councils or other political groups to access the money locked up in that property. And no matter how many "in perpetuity" and "for future generations" clauses are imbedded in an agreement, it can still be compromised or upended by private interests with clever lawyers. There is great sadness in seeing land that was fought for by dedicated activists or generously donated by passionate philanthropists return to private or commercial interests. That's why I started a program at RRI called Defense of Place.

The program is based on the legal doctrine of the Public Trust, a pillar of our democracy. It states that some public assets are so valuable to the health and equity of society that they cannot be sold or given away. Instead, they must be managed in trust for the benefit of all people for all time.

Defense of Place offers RRI's experience and support to citizens around the nation in their crusade to protect public lands – parks, open spaces, preserves, and places of environmental significance – that are threatened by economic or political interests. Any betrayal or manipulation of the donor's or public's intent represents a loss of trust in all contracts that bind generations to a common heritage of land stewardship.

The question is one of permanence and value above and beyond money. Defense of Place is committed to making sure our protected lands stay that way. Forever means forever.

RRI has also taken on the defense of a place so important and imperiled it needed a program all its own – Restore Point Reyes Seashore. Many of us worked for decades to preserve the unique beauty and biodiversity of this national treasure north of San Francisco. We negotiated in good faith with local cattle ranchers who agreed to sell their coastal grazing land to the US Park Service at a fair price. We even agreed to lease back the land the ranchers had sold for a twenty-five-year period or for as long as they lived. Thanks to effective public relations and hefty political contributions, those leases have been continually renewed for more than half a century. Today, the park's land and water are dangerously polluted and degraded by cattle grazing, menacing the health and habitat of hundreds of native plant and animal species, including the magnificent native Tule elk. The ranchers have grown even more politically powerful and are attempting to grab more land, expand their operations, and kill off any Tule elk that get in their way.

There is no reason why private businesses should benefit from taxpayer-supported parkland intended to preserve wilderness areas for public enjoyment. Yet we must continue to fight for what rightfully belongs to the American people. As always, we will persist, petition, protest, and sue for as long as it takes to wrest Point Reyes National Seashore – and all US national parks and public lands – from commercial usurpation.

One of our most enlightening and foresighted programs was conceived, researched, and realized by someone who doesn't even work at RRI – my longtime friend, Steve Steinhour, whose intelligence and work ethic contributed so much to my efforts in Sacramento. Long before the climate crisis was on the average person's radar, Steve became concerned with how the extreme weather conditions it causes affect our communities and lives. He dived into the subject with great energy and commitment, creating a unique information source called the RRI Extreme Weather Community Resilience Project. It recounts experiences of communities in the United States that are adapting to the increasing frequency of extreme weather, so other places can learn from them. The project also explores weather and climate impacts in other nations to help clarify how the warming atmosphere may change the physical, economic, political, and cultural realities of our world.

Steve's insights have had quite an impact on how we view our own programs at RRI – the climate-changing effects of cattle ranching and rising sea levels at Point Reyes, the importance of reducing methane emissions in rice farming, the implications of megadrought conditions in California and the West.

The RRI program closest to my heart is Forces of Nature: Environmental Elders Speak. I started it for two reasons. First, I had long been envious of the book-lined libraries at law firms, filled with case studies and legal precedents. I thought it would be invaluable for environmentalists to have the same thing, a single archive of environmental history and knowledge they could easily refer to.

My second reason came from a great loss: the death of my friend and inspiration, David Brower, one of the most effective environmental voices in the world. He had shared his wisdom with me over many years, and when he was gone, there was a vacuum. Sadly, I had never documented his words, and I vowed I would not let that happen again.

So Forces of Nature became a unique kind of archive, one that chronicles the achievements and lessons of the great pioneers of the modern environmental movement. And instead of filling a library with large tomes, my team and I chose a more contemporary and accessible medium – online video interviews. Working with talented professionals, Lauren Veen, Lynn Alexander, and Vincent Tremblay – and trusted regulars, Deb, Chance, Elizabeth Baker, and Peggy Lauer – we started conducting interviews in 2013. By 2019, we had completed more than 160. With more than 600,000 visitors from 147 countries, the website is clearly filling the need I had hoped it would.

When I began the project, I expected the interviews would be packed with practical, replicable environmental experiences and strategies, and they are. What I didn't expect was just how moving it would be for me personally. Forces of Nature gave me the opportunity to reconnect with wonderful colleagues and friends, and to revisit the enormous significance of what they had accomplished. Reclaiming contaminated rivers and bays. Restoring wetlands. Stopping dam construction. Banning dangerous chemicals and pesticides. Planting and replanting forests. Introducing children to nature. And most important, inspiring and mobilizing thousands of others to do the same. One other thing I noticed over the course of 167 interviews: as different as each interviewee is from the other, they all share one quality. They are never deterred. Never.

Along with the programs under our own roof, RRI sponsors a number of worthwhile nonprofits whose efforts expand our own. Some are ardent defenders of beautiful public lands threatened with encroachment like Californians for Western Wilderness (CalUWild), whose coordinator, attorney Michael Painter, was a former RRI staff member; and the Washoe Meadows Community directed by Lynne Paulson. Daniel Heagerty oversees two organizations sponsored by RRI: the Granite Chief Wilderness Protection League and Generation Our Climate, a high-powered, student-run, climate action group. In addition, the Public Trust Alliance, led by executive director Michael Warburton, uses the legal framework provided by the Public Trust Doctrine to protect and defend our natural resources for future generations.

Looking back at these thirty-five years, I'd say it's been a good way to have lived nearly half my life so far. My children grew to adulthood here in Mill Valley. I'm proud that they are vigilant protectors of the environment and equally proud that they have chosen career paths of their own. My daughter, Megan, has her own thriving, community-minded pet store and works for pet adoption. My son, Tyler, has developed and managed technology systems for many of the Bay Area's leading startups. His wife, Jill Kaufman, is part of a company innovating environmentally sustainable products. Their two sons, Miles and Bay, are both college students and exploring the possibility of careers in law and politics. My wife, Sue, has had a lifelong commitment to causes that promote social and environmental justice, and she has spent countless hours volunteering on their behalf. Not only that, she became a rower in her fifties, joining a group of women who went on to win a world championship in their class. One of the best parts of my life is our family dinner on Sunday nights. We have a tradition of serving protein I've killed on hunting and fishing trips – duck, salmon, wild boar, or elk. And while Megan is a vegetarian, she makes an exception at our dinners, knowing the animals lived their lives in the wild and were killed according to laws that protect their species.

At RRI, I've had the chance to spend my time working with people I like and admire. I've been able to do what I love to do – thinking up ideas and giving others the opportunity to do the same. I've been a stepping stone for many – probably dozens – who have gone on to build outstanding environmental organizations of their own. And I've traveled the world, engaging with environmentalists in Beijing, Rio, the Netherlands, Nairobi, Norway, New Zealand, Singapore – and Moscow, St. Petersburg, and Lake Baikal when the Soviet Union was on its last legs. I even got to attend my friend Wangari Maathai's Nobel Peace Prize ceremony and banquet in Oslo, a moving and elegant event.

The Award of a Lifetime

I've been fortunate to receive many environmental honors and awards in my life. To be frank, as much as I appreciate the gesture, awards have never meant much to me. Accomplishing a hard-fought win for the environment is where my satisfaction lies. Often, when an award comes my way, I'm already on to the next battle and can barely remember what the honor is for. It's the doing that matters to me.

There is one exception to that rule – winning the 2001 United Nations Environment Programme (UNEP) Sasakawa Prize. The experience was both exhilarating and humbling.

UNEP had been part of my life since my 1972 trip to Stockholm when I jumped over the hedges and sneaked into the conference

My delightful friend and colleague, Nobel Peace Prize winner Wangari Maathai.

where the agency was first being planned (before that, actually, as a number of us had been campaigning to establish a UN environmental agency for several years). I was still at The Nature Conservancy then, and the International Chamber of Commerce was conspiring to make sure the new organization would be environmental in name only. Fortunately, the fuss my friends and I caused in Stockholm stopped their plan in its tracks.

It was through UNEP that I met and became close friends and colleagues with Wangari Maathai, the brilliant and courageous founder of the Kenyan Green Belt Movement. In Chapter Six, I describe how Wangari put her life on the line to protect and restore Kenya's natural environment, then won the Nobel Peace Prize for her efforts. When I finally started my own nonprofit, I was pleased to help her launch the Green Belt Movement International to expand her tree-planting crusade worldwide. Apart from the obvious value of the project, I had an ulterior motive for taking it on. I hoped it would keep Wangari's name in the news and protect her from getting killed by enemies in the corrupt Kenyan government. It seemed to do some good, thank goodness.

In 2001, without my knowledge, Wangari and my wily staff submitted my name for UNEP's prestigious annual prize. UNEP defined the $200,000 prize as "recognizing outstanding individuals and organizations for their significant contribution to the protection of the environment and the promotion of sustainable development."

A few months later, I received something in the mail from UNEP. I was still in the dark, so it never occurred to me that the letter could have anything to do with the Sasakawa Prize. For days, I left it unopened on my desk. After a while, someone from UNEP gave me a call to see if I had received the letter. That's when I found out that I had,

Receiving the 2001 United Nations Sasakawa Environment Prize, the most memorable night of my professional life.

in fact, been given the award and needed to set a date for a celebration in New York.

That astonishing phone call was only the first of many surprises. For starters, my staff had secretly arranged to have a hundred friends from the Bay Area fly to New York to attend the ceremony. The first I knew about it was when I walked into the UN lobby shortly before the presentation was to begin and saw them all there, dressed to the nines and enjoying cocktails and hors d'oeuvres. I felt almost delirious with joy that the people who meant the most to me could share in this once-in-a-lifetime experience.

At the cocktail party, I found out about the next surprise. I was handed a message from Martha Lyddon, one of my most loyal and generous supporters. She was sorry she couldn't attend the ceremony, but she had booked a large ballroom at a hotel next to the United Nations and arranged a post-ceremony banquet for everyone to enjoy – including the UN employees who were not usually invited to such events.

The ceremony was something to behold. Queen Noor of Jordan gave an eloquent speech. Klaus Töpfer, executive director of UNEP, read a congratulatory message from the UN secretary general, Kofi Annan, an honor I still can't quite believe I actually received.

His message begins:

> It gives me great pleasure to recognize the achievements of this year's winner of the United Nations Environment Programme Sasakawa Environment Prize. Mr. Huey Johnson, whose unflinching support of environmental causes for more than 40 years has won global admiration....

It goes on to say:

> The quality of human life ultimately depends on the quality of the environments in which humans live, and

the ability of those environments to provide food, shelter, jobs and sustenance. If we want to improve standards of living, we will undermine our own efforts if what we do in the name of peace and progress damages the earth's precious ecosystems and natural resources. Today, the warnings of ecological crisis are loud and clear; we must respond more aggressively, more promptly and more globally than we have to date.

One person who has taken this challenge to heart is this year's winner of the Sasakawa Prize. Huey Johnson recognized long ago the inequitable distribution of the earth's natural resources, and devoted his career to developing better ways to understand and manage those resources. He has understood that responsibility for conservation and environmental protection lies with every single member of the human community. And he has founded numerous organizations – the Trust for Public Land, the Resource Renewal Institute, the Green Belt Movement International, the Grand Canyon Trust and the Environmental Liaison Centre International – that have developed policies and crafted programmes that have been emulated throughout the world.

His commitment to making this world a better place is an inspiration to us all. It is with great pleasure that I call upon the Executive Director of UNEP, and the Director of International Affairs of the Nippon Foundation, to present, on my behalf, the 2001 UNEP Sasakawa Environment Prize to Mr. Huey Johnson. Let us congratulate him for all he has done to advance sustainable development and to cultivate a new ethic of stewardship for the earth and its inhabitants.

The surprises kept coming. The English nobleman who headed the Sasakawa Prize selection committee – his name escapes me – gave a speech that went into all the politics leading up to my receiving the award. It was as if he were talking to his wife, not an audience of dignitaries and diplomats. Still, the story was interesting. The European men on the committee, including the Englishman telling the story, did not want me to win the award at all. They went so far as to blackball my submission after I had already received the majority of the committee votes. It's hard for me to be objective, but I suppose their argument had merit; they believed it was high time a person from the developing world be chosen as the recipient. Ironically, it was the women of color on the committee, including Wangari, who fought for me and refused to back down. They believed my work greatly benefited the developing world, and that was more important than my nationality, race, or gender. They won the day and I won the prize.

As I said, I'm not big on awards, but this was a day to remember. I look back on it all with joy and gratitude.

I could fill countless more pages with stories about my life at Resource Renewal Institute. I'll add just one more because it embodies much of what RRI means to me – wilderness, friendship, generosity, and, as always, a good dose of salesmanship.

I was elk hunting in the Colorado Rockies near the town of Steamboat Springs. I had two companions, my longtime friend and RRI board member Henry Corning, and my grown son, Tyler. Each day we would hike up the nearby mountain for several miles, and when we reached the timberline, split up and hunt elk until late afternoon. Then we'd find one another and hike back down the mountain.

It had been snowing and the snow was getting deep. By the third day, hiking up the mountain was becoming something of an ordeal. Still, we managed to make it up to the timberline. As usual, we carried packs with emergency gear and food just in case it stormed so hard we couldn't hike back and would have to stay overnight on the mountain. For some reason, I decided it would be nice to join up with my partners and eat our lunch together. There is nothing better than hot soup after a morning of wading in snow above your knees. I saw them in the distance and flagged them down. We built a small fire and sat down to talk and enjoy lunch.

After our lunch, we put out the fire and parted ways. I started walking into a wooded area. Not fifty yards into the timber from our lunch spot, I noticed a huge bull elk. He was lying there as if he had been listening to our conversation. He stood up and stared at me.

An elk rifle has a telescope, and the crosshairs in mine were on him for a killing shot. We both stood still, and I studied him. He looked rather thin to be living at that rugged elevation. I assumed he was what is known as a herd bull. I stood aiming for a long while as an argument occupied my mind. I realized I didn't want to kill the father of a lot of the elk who lived on that mountain.

Male elk fight to see who gets to live and breed with a herd of cows, usually twelve to twenty in number. The losers in the battling maintain a comfortable distance from the alpha bull, but once in a while they will bugle and challenge him to a duel. The alpha male is also challenged repeatedly by junior bulls who then retreat and gather up their courage for another try.

As the challengers line up, the alpha has little time to eat or drink. He will often tire, weaken, and die in the cold winter after breeding season ends. So it was with the elk before me, terribly thin and hiding in a high place.

Although I had decided not to shoot this magnificent animal, I called out to Henry who was still within earshot. This might have been the only elk we would see that week, and I wanted to give Henry a chance. He didn't hear me, and the elk ran away.

It snowed heavily that night, and the next morning, my partners loaded up and trudged back up the mountain, one breaking trail and the other following in his tracks. At that moment, I decided to stay behind in the flat areas by the road. Walking toward where our truck was parked, I passed a ranch house. Just then, someone came out of the house, and I asked him if I could hunt in the forest behind his home. He said, "Sure, but keep your eyes open for cows. A lot of them have gone missing in this heavy snow." He was getting into his truck to try and track them down. On the spur of the moment, I offered to help him look for his cows. He smiled and opened the passenger side door.

We exchanged names. His was Jay Fetcher. We did a lot of talking and found a cow now and again. He said he had already heard of me because he was looking into ways his ranch could become public meadowland. He had already met with The Nature Conservancy and The Trust For Public Land, but he felt they were competing with each other to get his ranch, and it made him uncomfortable.

I advised him to forget both TNC and TPL, then sold him on the idea of starting his own land-saving organization and doing things his way. It must have been a good piece of advice because, before long, he had turned himself into a world-class land saver. He started an organization called the Colorado Cattlemen's Agricultural Land Trust and convinced his fellow ranchers to preserve the development rights to thousand of acres of ranchlands. CCALT became the first group of cattle ranchers to organize themselves for land saving, an idea that has spread to several other states. When I spoke with Jay recently, he said he and his fellow cattle ranchers had transferred more than five hundred thousand acres to meadowlands in the public trust. He kindly credited me with having inspired him that day in his truck. I see it as a clear example of the power of salesmanship – both mine and his. Not to mention a bit of good luck that I had decided against elk hunting on that snowy day.

Months later, Henry wrote a beautiful poem about our elk-hunting trip that yielded no elk. Its final two lines are etched in my memory:

The bull listens and waits, and after I move away reveals himself
To one wise enough to spare a deity.

CONCLUSION

Lesson Learned | Follow Your Nature

Leonardo da Vinci described water as the driving force of all nature. But until my friend Steve Steinhour mentioned it to me, I never realized that water is the driving force of *my* nature, and has been my whole life. As children of six and seven, my two best friends and I whiled away our summers at a stream near home, fishing by day and camping by night. In winter, my father and I loved to ice fish on one or another of Michigan's eleven thousand lakes. And during college, I'd get off work around midnight and head straight to the local river for a swim.

When I graduated from Western Michigan University in 1956, the economy was booming, and I had my pick of several attractive, well-paying jobs. Characteristically, I decided to take a short fishing trip to think over my options. On my drive to the lake, I stopped in Grand Rapids for a quick walk along the Grand River.

The Grand once roared with dramatic cascades, dropping eighteen feet within a short distance and giving the city of Grand Rapids its name. As the town grew, its citizens built a series of fourteen dams, flooding the rapids to make way for grain mills and sawmills that would connect to river transportation. In the twentieth century, Grand Rapids became an auto industry center and its river a dump for factory waste and pollution.

As I walked along the Grand, it stank to high heaven and glistened with oil and grease. At that moment, I made a snap decision: I was not going to spend my life in a state where people didn't give a damn – about nature, about pollution, about the place they called home. That night, I turned my back on my lucrative job offers, packed up my car, and left Michigan for good. Water – in this case filthy, poisonous water – was the driving force that led me to a better life.

As my life took its course, water became both the focus of my work and the way I escaped from work. At The Nature Conservancy, I spent my weekdays figuring out ways to preserve Hawaii's Seven Sacred Pools or the California coastline, and my weekends fly-fishing, river rafting, or crabbing on San Francisco Bay. As secretary of resources, I'd get away from the frustrations of battling California's all-powerful water lobby by swinging from a rope and jumping into the Sacramento River. And when the time came for me to find money for one of my nonprofits, I often raised funds in a rainy wetland duck blind I shared with generous CEOs and philanthropists.

I'm not the first one to point out that there is a lot to be learned from water. But that doesn't make it any less true. To me, it embodies everything an environmentalist should be. Relentless, persistent, tenacious. Willing to find progress in a small drop of change. Water has the power to transform the earth, carving canyons and making deserts bloom. I believe the environmental movement has the power to transform human beings – through education, persuasion, and sometimes creative conflict – so we can live in harmony with the earth.

It turns out that water has another miraculous power I wasn't aware of – the power to give my book a happy ending. Just a few months before finishing up this last chapter, I received some news I'd been hoping for for more than twenty-five years. Back then, California was experiencing a terrible drought. Nobody seemed to be doing anything to help the suffering wildlife, so I drove up to Lake Tahoe to investigate for myself. I was shocked by what I saw. The lake level was so low, piers and boats were stranded far from the lake's edge. Where streams normally drained into the lake, there was cracked mud that smelled of dead things. A few weeks later, I traveled to Montana to fish in a stream prized for its trout. Like Tahoe, it was dangerously parched. To make matters worse, private farmers were arrogating water from the public stream to irrigate their crops and soak hay.

Deeply affected by what I saw, I decided something had to be done. I thought of how I had helped create the profession of nonprofit land saving at The Nature Conservancy and wondered if I could do something similar to save water for the public good. When I got back from Montana, I did just that, establishing the first water heritage trust as part of the Resource Renewal Institute.

I started calling on government agencies that managed water assets to see if any of their contacts might want to donate or sell their water rights to the public trust. Sure enough, I got a call back from a federal water specialist. He knew an elderly couple who had sold their old ranch beside beautiful Butte Creek but still owned a portion of the water rights. By a stroke of good luck, it turned out that Butte Creek harbored Spring Run Chinook, the rarest run of salmon in all of California.

As I used to do back in my Nature Conservancy days, I spent quite a lot of time getting to know the couple personally, educating them on how important their donation would be to the environment, wildlife, and people of California. We talked it over a number of times. I made sure they understood that the whole idea of a public water trust was in its infancy, and I didn't have any money to make the purchase outright. They agreed to let me option the rights for an agreed-upon price. (Neither they nor I knew then that we would have to extend the option at least a half a dozen times.)

Option in hand, I went directly to various government departments in charge of salmon management to see if they would be interested in buying the rights. They were as excited by the prospect as I was – and just as unable to pay for it.

Because everything about obtaining and legally protecting water instream for fish and wildlife was brand-new, my tale of Butte Creek is long and complicated, involving state and federal agencies, water law, court hearings, and reclamation appropriation bills. For years, RRI staff made headway, then hit brick walls. But when Deb Moskowitz joined us, it became her holy grail. She waded into the morass, spending nearly a decade taking the process forward step by complicated step. With the help of several knowledgeable members of the "RRI cadre" – especially veteran water attorney Alan Lilly who patiently stayed with the project year in and year out – she studied water policies, researched water master reports, and wrote to state officials and federal bureaucrats again and again.

Many times, we were tempted to let the whole thing go, but through it all, one simple truth stood out. The state of California, and the environmentally minded people at the Department of Fish and Wildlife, saw the urgent need to permanently protect Butte Creek water, if only there was a way to make it happen.

And then it happened. In 2018, the wise people of California passed Proposition 68, authorizing $4 billion for "parks, environment and water." Prop 68 – along with a 2014 bond measure that included appropriations for "ecosystem and watershed protection and restoration" – made funding available for a project like ours.

Our work started all over again. We had to make a strong and compelling case that, of all the requests received by the state of California, Butte Creek water rights deserved to be funded. More forms, applications, and letters of recommendation followed. The first time, we were turned down. And the second. Then in March of 2020, after twenty-five years, our request was approved. At last, RRI's water rights will be permanently dedicated to this lovely and critical stretch of water. Home to a rare species of California salmon, it will exist forever in its natural state.

The victory of Butte Creek, so long in coming, gives me hope that we environmentalists will meet the staggering challenges facing us today. So does Grand Rapids, the place I left in disgust all those years ago. The city is in the midst of removing its dams and restoring the glorious rapids that no living soul has ever laid eyes on. In the next few years, lake sturgeon and snuff box mussels are expected to make their return, along with rafters and hikers and fishermen.

Hope, like water, springs eternal.

| *Marincello a half century later. Photo by Gary Yost.*

Bibliography

Andrus, Cecil D., and Joel Connelly. *Politics Western Style.* Seattle: Sasquatch Books, 1998.

Baron, Robert C. *Heaven & Nature Sing: Land, Wilderness, and Writers.* Golden, CO: Fulcrum, 2009.

"Battles Won." *Life Magazine,* Time-Life Publications, July 4, 1970.

Borden, Charles A. *Sea Quest: Global Blue Water Adventuring in Small Craft.* Philadelphia: Macrae Smith, 1967.

Brand, Stewart, ed. *CoEvolution Quarterly.* Sausalito, CA, 1974.

Brand, Stewart, ed. *Whole Earth Catalog.* Menlo Park, CA, 1968.

Cahill, Russell. *Tales from the Parks: My Adventures as a Park Ranger.* Seattle: Amazon, CreateSpace Independent Publishing Platform, 2016.

Congressional Record. RARE II. https://uscode.house.gov/statutes/pl/98/425.pdf.

Conrad, Joseph. *Lord Jim.* New York: Modern Library, 1931.

Forces of Nature: Environmental Elders Speak. Mill Valley, CA: Resource Renewal Institute, 2012. theforcesofnature.com.

Frommer, Arthur. *Europe on Five Dollars a Day.* New York: Frommer's, 1957.

Hart, John. "How Grit and Grace Saved Marincello." Berkeley, CA: *Bay Nature,* July–September, 2003.

Ickes, Harold L. *The Secret Diaries of Harold L. Ickes* (*Volume I: The First Thousand Days 1933–1936*, 1953; *Volume II: The Inside Struggle 1936–1939*, 1954; *Volume III: The Lowering Clouds 1939–1941*, 1954). New York: Simon & Schuster.

Johnson, Huey D. *Green Plans: Greenprint for Sustainability.* Lincoln: University of Nebraska Press, 1995.

Kelly, Nancy, and Kenji Yamamoto. *Rebels with a Cause.* KRCB, North Bay Public Media, 2012.

Kirk, Andrew G. *Counterculture Green: The Whole Earth Catalog and American Environmentalism.* Lawrence: University Press of Kansas, 2007.

Leopold, Aldo. *A Sand County Almanac.* Oxford, UK: Oxford University Press, 1949.

Seuss, Dr. (Theodor Seuss Geisel). *The Lorax.* New York: Random House, 1971.

Ward, Barbara. *Spaceship Earth* (George B. Pegram Lectures). New York: Columbia University Press, 1966.

Ward, Barbara, and René Dubos. *Only One Earth: The Care and Maintenance of a Small Planet.* New York: W. W. Norton, 1972.

Wenkham, Robert. *Maui, the Last Hawaiian Place.* New York: Friends of the Earth, 1971.

Wood, Samuel E., and Alfred E. Heller. *California Going, Going: Our State's Struggle to Remain Beautiful and Productive.* Sacramento: California Tomorrow, 1962.

Index

Adams, Ansel, ix, 171, 172
Admiralty Island, 45
Alaska, 40, 42, 43, 46, 48, 49, 50, 125, 211
Alaska Department of Fish and Game (ADFG), 42
Alexander, Lynn, 241, 255
American Society of Landscape Architects, 145
Andrus, Cecil, 152, 211, 212, 213, 214
Angel, John, 242
Archbald, Greg, 126, 133
Arnold, Elting, 119
Arnold, Herb, 101
A Sand County Almanac, xiii, xv, 38, 39, 41, 152, 153
Audubon Society, 70, 184, 198
Australia, 28, 29, 34

Baker, Elizabeth, Baker, 239, 241, 255

Baker, Richard, 85, 149
Bangkok, 30, 225
Bank of America, 135, 141, 198
Barcelona, 31, 32, 40
Barzaghi, Jacques, 156
bears, xi, 44, 45, 46, 47, 214
Behr, Peter, 103, 104, 179, 210
birds, 28, 58, 59, 66, 72, 76, 77, 92, 103, 137, 182, 184, 190, 191, 210, 222, 249
Black Panthers, 140, 212
Boardman, Walter, 112
Boise, 12, 51,
Boise Cascade, 221
Bolinas Lagoon, 74, 102, 105, 134, 147
Bolle, Dr. Arnold, 160
Borden, Charles, 79, 80, 81, 82, 83, 114, 115,
Borden, Eleanor, 80, 81, 83, 115
Brand, Stewart, 135, 137, 147, 169

British National Trust (BNT), 116, 117, 118, 121, 128
Brower, David, ix, 20, 71, 148, 149, 235, 255, 282
Brown, Jerry, ix, 3, 126, 142, 148, 149, 150, 153, 157, 158, 196, 197, 199, 203, 208, 216, 227, 239, 283
Bryan, Bill, 137, 220
Burton, Phillip, 201, 204, 207, 208
Butte Creek, 268, 269, 270
Byrne, Albert, 119

Cahill, Russell, 116
California Agricultural Aircraft Association, 175
California Cattlemen's Association, 179
California Department of Fish and Game, 38
California Department of Resources, 3, 194
California Secretary of Resources, x, 147, 210
California Waterfowl Association, 171
California Water Trust Network (CWTN), 249
Californians for Western Wilderness, 256
Cameron, Francis Baldwin, 66, 67, 68, 70

Canada, 229, 230, 236
Canelo Hills Cienega Preserve, 57, 283
Carnegie, Andrew, 2, 121, 128
Carroll, Jennifer, 232
Charles, Prince, 151
Chicago, 12, 16, 17, 53, 75, 129, 130, 143, 169
Chinn, Jean Wetzel, 241
Chouinard, Yvon, 220, 249
Coeur d'Alene, 138, 139
Collins, B. T., 173, 174
Colorado Cattlemen's Agricultural Land Trust (CCALT), 264
Conservation Fund, 125, 126
Cooley, George, 112
Corning, Henry, 262
Costa, Steve, 126, 146
cranes, xv, 190, 191, 192
Cravalho, Elmer, 78
Cutrano, Chance, ii, xv, 244

Davis, Gray, 157
Defense of Place, 241, 253
DeJongh, Paul, 233
Denton, Jan, 157, 164, 179
Denver, 12, 13, 15, 17, 50, 91
Deukmejian, George, 203, 204, 216
Deutsch, Barbara, xi, xiii
Devlin, Shawn, 251
Douglas, William O., ix, 113, 115

Dreyfus, Helen "Babby", 222
Dubois, Mark, 171
Dubos, René, 106
Ducks Unlimited, 171
Duke, Doris, 76

Erdman, Chris, 220
Export Excess, 224

Federal Land Policy and Management Act (FLPMA), 88
Ferguson, Doug, 96, 99, 101
Fetcher, Jay, 264
Fish in the Fields, 249, 250, 251, 252, 284
Fookes, Tom, 233
Foote, Sy, 123
Forces of Nature: Environmental Elders Speak, 255
Ford Foundation, 112, 132
Fort Cronkhite, 217, 225
Freemasons, 24
Fuller Brush, 9

Gellert, Annette, 220
Generation Our Climate, 256
Gerard, Jenny, 129
Glacier Bay National Park, 42
Golden Gate National Recreation Area (GGNRA), 99, 100, 127, 134, 208, 217, 283
Golden Trout Wilderness, 183

Goodspeed, Richard, 122, 123
Graalman, Nancy, 243
Grand Canyon Trust (GCT), 226, 261, 283
Grand River, 266
Grand Traverse Regional Land Conservancy (GTRLC), 117
Granite Chief Wilderness Protection League, 256
Great Depression, 37, 152
Greenberg, Phil, 188
Green Gulch Ranch, 57
Green Party, 143
Greenplanning, xx, 129, 229, 230, 232, 233, 234, 235, 236, 237, 242, 249, 250
Green Plans, 229, 231, 232, 233, 234, 235, 242, 284
Griffin, Marty, 74, 99, 103, 104
Gulf Oil, 60, 94, 97, 99

Haleakala National Park Volcano, 65
Hamburg, 34, 35
Hammond, Rich, 164, 189
Hayes Valley, 144, 146
Hearst, Will, 227
Heller, Alf, 219

IBM, 15, 123, 198
Ickes, Harold LeClair, 151, 152, 156

Inholdings, 138, 139
Investing for Prosperity (IFP), 194, 196, 199, 214, 228, 237, 238

Jewett, George, 75
Johnson, Duane, 2, 3
Johnson, Marilyn, 3, xix
Johnson, Megan, iii, 151, 153, 256, 257
Johnson, Rebecca, 3
Johnson, Sue, iii, xv, 40, 53, 80, 153, 155, 256, 284
Johnson, Tyler, iii, xix, 153, 242
Jordan, Allison, 239

Kanaha Pond State Wildlife Sanctuary, 66
Kaufman, Jill, 256
Kent, Mrs., 103, 104, 147
Kettenhoffen, "Ket", 103, 104
Kibbey, Jason, 241
Kirk, Andrew G., 136
Koshland, Daniel, 144, 146
Koshland Park, 144, 145
Krutch, Joseph Wood, 61
Kuala Lumpur, 29, 30

Lake Tahoe, x, 38, 40, 42, 267
Land, Edwin, 84
Lauer, Peggy, 220, 227, 255
League of Women Voters, 197

Leopold, Aldo, x, xiii, xv, 38, 41, 52, 61, 110, 152, 245
Leslie, Leslie, 220, 243
Levinsohn, John, 242
Lincoln, Abraham, 7, 119
Lindbergh, Charles, 71, 74, 76
Lithman, Alan, 136
Livermore, Caroline, 89
Livermore, Putnam, 89, 90, 122, 124, 129
lobbyists, 77, 152, 170, 174, 175, 176, 177, 178, 180, 186, 198, 211
London, 33
London, Jack, 50
Lovitt, Ron, 221
Lyddon, Dorothy, 219
Lyons, Bill, 155, 194

Maathai, Wangari, ix, 92, 110, 224, 238, 257, 258
Macy, Tom, 125
Marckwald, Kirk, 157, 164, 165, 213
Marcus, Vera, 163, 179, 211, 214
Marincello, 94, 95, 96, 97, 98, 99, 100, 103, 217, 270
Marin Headlands, 94, 99, 100, 147, 217, 283
Martin, Paul, 61
May, Cordelia Scaife, 60
McCaughey, Hamilton, 70, 79, 93

McLaughlin, Sylvia, 220, 221
Mead, Margaret, ix, 108, 109
Mellon, Andrew, 60
Mexico, 58, 89, 234, 235, 243
Michigan State University (MSU), 7, 8
Mills, Stephanie, 108, 111
Milton, Bill, 90
Minton, Jonas, 212, 252
Mono Lake, 184
Montoya, Joseph B., 157
Morrill Act, 7
Moskowitz, Deborah, xvi, 243, 244
Mother Jones, 227

National Park Service, 78, 99, 134
Nelson, Daisy, 70
Nelson, John, 127, 128, 129
Nevada Outdoor Recreation Association (NORA), 87, 88, 89
New Delhi, 24
New York, 5, 12, 14, 16, 61, 67, 75, 76, 77, 101, 112, 128, 133, 226, 260
New Zealand, xvii, 23, 25, 26, 27, 28, 34, 79, 231, 233, 235, 239, 257
Noor, Queen, 151, 260
nuclear power, 149, 187, 188

Oakland, 123, 126, 135, 141, 143, 189, 239
O'Keeffe, Susie, 232
open space, x, 34, 55, 56, 64, 110, 114, 117, 120, 121, 122, 126, 130, 133, 134, 135, 141, 142, 143, 166, 167, 190, 227, 231, 241, 253
Ordway, Gil, 222

Painter, Mike, 256
Patagonia, 220, 249
Patagonia-Sonoita Creek Oasis, 57, 60, 65
Peet, Crayton, 123
Penang, 29, 30
Point Foundation, 136, 137, 138, 139
Point Reyes, 100, 102, 142, 208, 217, 238, 254, 283
Polaroid, 84
Praetzel, Bob, 96, 99
Price, Marilyn, 241

RCA, 100, 101, 102
Redford, Robert, 240
Reed, Scott, 138
Reid, Lewis, 123
Resource Renewal Institute (RRI), x, 129, 216, 219, 220, 228, 230, 233, 234,

236, 237, 239, 241–249,
252–257, 262, 269, 283
Restore Point Reyes Seashore,
253
Rio Earth Summit, 239
Rockefeller, Laurance, 67, 68, 70,
72, 73, 76, 93, 108
Rockefeller, Mary, 76
Rogers, Don, 214
Rollins, Darcy, 232
Rome, 32, 38
Ronstadt, Linda, 62, 169
Rosen, Marty, 96
Royko, Mike, 169
Russell, Andy, 238

Sacramento, 38, 124, 131, 150,
152, 153, 156, 157, 161,
164, 167, 168, 176, 177,
180, 183, 187, 188, 192,
193, 197, 203, 212, 217,
218, 221, 229, 238, 240,
247, 252, 254, 284
Sacramento River, 192, 267
salmon, 42–46, 51, 173, 222,
249, 250, 257, 268, 270
San Francisco, 13, 19, 54, 57, 60,
61, 67, 72, 75, 77, 79, 81,
82, 89, 90, 91, 94, 100, 101,
102, 109, 110, 113, 115,
119, 120, 134, 135, 144,
145, 155, 181, 212, 214,

220, 221, 240, 253, 267,
283
San Francisco Examiner, 105,
227, 238, 244
San Francisco Zen Center, 34,
87, 127, 146, 149, 207
Schifferle, Patty, 170
Schwartz, Debra, 99
Second Roadless Area Review and
Evolution (RARE II), 194,
200, 201, 203, 206–209,
214, 252
Seiberling, John, 205
Seton, Joe, 232
Seven Sacred Pools, ix, xx, 34, 58,
65, 68, 69, 70, 71, 74, 108,
167, 215, 219, 267, 283
Shanks, Bern, 221, 245
Sierra Club, 148, 149, 170, 198,
200
Silberman, Richard, 196, 222
Sips, Herman, 233
Skov, John, 242
Sladen, Jocelyn Alexander, 222
Sonoran Desert, 57, 59
Spindrift Point, 80, 82, 83, 85, 114
Steele, Ted, 57, 58, 65, 67, 71,
79, 89, 92, 93
Stegner, Wallace, ix
Stein, Pete, 129
Steinhour, Steve, 126, 131, 166,
254, 266

Stoddard, Charles, 158
St. Petersburg, 33, 257

The Nature Conservancy (TNC), 34, 54, 55, 56, 57, 67, 73, 77, 79, 82, 83, 85, 91, 92, 95, 103, 104, 112, 116, 122, 126, 129, 139, 147, 148, 149, 166, 167, 191, 215, 216, 217, 258, 264, 267, 268, 283
The New California Water Atlas (NCWA), 248
The New Renaissance Center, 218, 229
The Trust for Public Land (TPL), x, 34, 64, 120, 133, 143, 147, 148, 166, 167, 198, 215, 216, 245, 261, 264
Thompson, Richard, 133
Train, Russell, 107
trees, 39, 58, 59, 60, 100, 110, 137, 170, 189, 191
Tremblay, Vincent, 242, 255

UNEP Sasakawa Environment Prize, ix, 111, 257, 259, 260, 261, 262, 284
United Nations (UN), xx, 151, 260
United Nations Conference on the Environment, 106, 149

United Nations Environmental Programme (UNEP), xx
University of Arizona, 58, 61
University of California, Berkeley, 70, 73, 188, 241
University of California, Davis, 163, 186, 250
University of California, Santa Cruz (UCSC), 86, 87, 185, 186
University of Colorado, 50
University of Michigan, x, 53, 56, 137, 142, 151
University of Montana, 160, 251
US Bureau of Land Management (BLM), 88, 158, 159, 160, 161
US Forest Service, 39, 44, 182, 183, 201, 202, 203, 206, 208
Utah State University, 7, 52, 54

Valentine, Mark, 232
Van Zijst, Hans, 233
Veen, Lauren, 242, 255
Videmski, Laci, 242
Visking, 11, 12, 14, 15, 16, 18, 19, 20, 21, 28, 29, 50, 120

Wallin, Phil, 129
Wall Street Without Walls, 129
Walsh, Michaela, 224

wannigan, 42, 44, 49, 50
Warburton, Michael, 256
Ward, Barbara, 106
Washoe Meadows Community, 256
water interests, 184, 185, 247
Water Transfer Project, 248
Watson, Charles, 87
Watt, James, 211, 221, 225
Wax, Mel, 95
Weber, Marion French Rockefeller, xiv, 219, 223
Wenkham, Bob, 71, 76

Western Michigan University, 8, 11, 266
Wheelwright, George, 84, 86, 149
Whole Earth Catalog, 135, 140, 147
Wild and Scenic Rivers, 210, 213
Wilderness Act, 200, 201
Witter, Ann, 94
Wrigley family, 75

YWCA, 22, 30

"I should warn readers of an ever-present danger. Huey Johnson's ideas are quite contagious."

— David Brower
environmental pioneer,
founder of Friends of the Earth and
Earth Island Institute

About the Author

Long revered as a giant of environmentalism, Huey Johnson founded many of the movement's most important organizations – Trust for Public Land, Grand Canyon Trust, Green Belt International, and Resource Renewal Institute, a small but influential environmental think tank he led for more than thirty years.

He began his career as western regional director of The Nature Conservancy, where he gained recognition for saving such iconic places as Maui's Seven Sacred Pools, Montana's Bear Tooth Ranch, and Canelo Hills Cienega Preserve in Arizona. He also helped acquire the Marin headlands, just north of San Francisco, and other significant parcels along the California coast. These acquisitions became part of the Golden Gate National Recreation Area and Point Reyes National Seashore – and the subject of the 2012 documentary film, *Rebels with a Cause*.

In 1978, Huey joined the cabinet of California governor Jerry Brown as secretary of resources, where he transformed his department from one that enabled the exploitation of the state's natural resources by private industry into one supporting environmental sustainability and restoration. During his tenure, he was instrumental in preserving millions of acres of US public lands and protecting twelve hundred miles of wild and scenic rivers.

At Resource Renewal Institute, he built on his Sacramento experience by developing programs with global applicability, such as Green Plans and Fish in the Fields. In recognition of this groundbreaking work and his earlier achievements, he received numerous awards and honorary degrees, including the United Nations Environmental Programme Sasakawa Prize, environmentalism's highest honor.

Huey was known for his mentorship. At Resource Renewal Institute, he guided the careers of countless burgeoning environmentalists, many of whom went on to found their own consequential programs and organizations.

Huey lived in Mill Valley, California, with his wife, Sue, for nearly sixty years. He died in 2020, just a few weeks after completing the manuscript of his life story.